The Railways of Gosport

Dedication

To Reg Randell

ISBN 0 946184 25 9
September 1986

Typesetting by
Alphaset, 65A The Avenue,
Southampton SO1 2TA

Printed by
Netherwood Dalton & Co.
Bradley Mills, Huddersfield

THE RAILWAYS OF GOSPORT

INCLUDING

THE STOKES BAY AND LEE-ON-THE-SOLENT BRANCHES

by
Kevin Robertson

Contents

Introduction

There is today a peculiar fascination about the railways of Gosport. Something inspiring yet sad, intriguing yet fulfilled, something that sets the mind wandering back to a time when blood and sweat toiled to allow the iron rails to stretch out and reach the south coast – only the second railway to do so.

In its heyday there were no less than seven standard gauge passenger stations open, all within just a few miles radius. How they came to be built and why they failed to live up to the expectations of the speculators, the râison dêtre of this work. Perhaps a few of the questions may now be answered.

I first became acquainted with Gosport whilst living in the town in the mid 1970s. A colleague told me '. . . to find your way around you must know the old railways . . .', certainly a challenge I could hardly fail to take up, the results of which now appear a decade later.

Come with me then through the last 150 years, to a time when thousands trod the platforms of the stations, to war, work or pleasure to justify the navvies, contractors, surveyors, directors and shareholders who some say toiled without purpose. I believe not, for there was a time, not so long ago when the railway was an integral part of the life of the community.

Kevin Robertson

Area Map

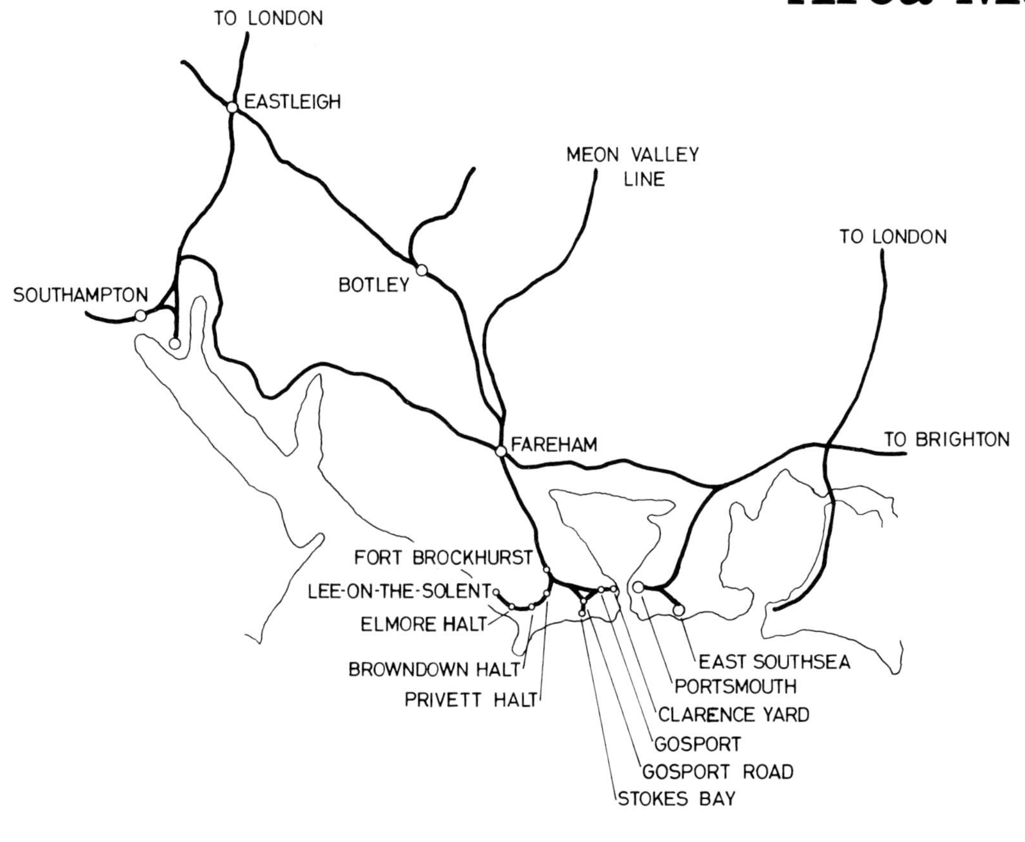

Chapter One

A Town in Hampshire

The town of Gosport, situated on the coast line of the south-east corner of Hampshire, occupies a unique and favoured position long recognised by the inhabitants of this small island of ours. For despite being bounded by water on three sides it is protected from the open sea, by Portsmouth, the Isle of Wight and the Solent, whilst still having easy access to the remainder of the county via the main Gosport to Fareham road.

But roads were far beyond the comprehension of the first known inhabitants of the area; these around 2000 BC when various stone age dwellings are known to have existed around the coast line. Little is then known of the vicinity until the 4th century AD when the town now known as Gosport was reported as being in the province of 'Meonwara', as the name implies this covered not only the Meon Valley but also large parts of the south coast from present day Dorset to Sussex.

The first real indications of the area developing towards its recognised 20th century state come in the years prior to the compilation of the Domesday book when tiny hamlets were reported at what are now Brockhurst and Elson. These were linked by Saxon tracks and today's roads often follow the same course.

With the compilation of the Domesday book on behalf of William I in 1086, twenty years after the Battle of Hastings, there is for the first time a clear record of the numerous tiny hamlets and dwellings of the area. 'Galemoor' later to become Gomer, Brocc-hyrst – for Brockhurst, Fortun or Fortune for Forton whilst Rowner retains the same naming. In addition there are certain names where rather more transition has occurred, 'Stoce' for Alverstoke and for the present day Elson 'Aethelswitheton' a connotation of Alfred's Queen, Ealhswith. Gosport itself has at various times been known as 'Godsport' or 'Gose-port', the latter indicating that geese were once sold in the market place. In addition certain members of the Royal Navy have adopted less polite references to the port.

The roads of the last century and looking east on what is now the A27 near to its junction with Redlands Lane at Fareham. In the background can just be seen the railway bridge carrying the Gosport branch. *Dennis Tillman collection*

Most of these early settlements were under the control of either the church or various Knights of the Manors. Some of these worthy individuals granting a modicum of freedom to their surfs and peasants compared with the usual slave labour adopted elsewhere. One of these was the large manor of Alverstoke, situated on the River Alver and taking within its boundaries the small Saxon or Jutish manor of Bedeham, (Bedenham). This latter locality was recorded in 1086 as consisting, 'one plough team and two cottage land labourers', having a total worth of 'half a hide' or 25/-.

The first known visit of a monarch to the area was in 1114 when Henry I visited Rowner, whilst later in 1277 it was the turn of Edward I on the occasion of granting the manor of Rowner to one Sir William LeBrun. This family has been associated with the area ever since and has several roads named after them. But in a general sense the area has changed little from its origins and is, with Gosport, a passive witness to the development of its near neighbour Portsmouth. Indeed Gosport was for several centuries destined to remain an obscure fishing village whilst its neighbour slowly developed into the premier port for the Royal Navy.

During the English civil war the towns of Gosport and Portsmouth were supporting different sides – Portsmouth the King and Gosport, parliament. This led to a blockade of Portsmouth by ships from Gosport and contributing in no small way to the surrender of what was one of the few royalist strongholds on the south coast. This may possibly be the time that the strategic value of Gosport was first realised for certainly from this time on there is reference to a slow build up of naval development in the area. Not, it should be pointed out, extending very far at this time; Lee-on-the-Solent for example which is but a couple of miles distant, was still a tiny hamlet and destined to remain so for some time to come. But in this way the walls surrounding the town were slowly raised, a feature to have such a profound influence on the progress of the railway later on. There were other defences too, numerous forts erected around the coast line intended to be a line of defence against the French.

This of course was the period of the Napoleonic wars and the town developed with the navy and so became host to large numbers of captured French prisoners. These unfortunates were housed in what can only be described as terrible conditions, mainly at the old Forton Hospital, converted to a prison for the purpose and seemingly lacking in many of the basic care facilities. Many of these wretched individuals were to attempt to escape, some successfully, whilst others were hanged when caught, to be buried in vast graves within the St. Vincent area.

The town too had its share of death, especially in 1664 when plague was rampant. Many inhabitants moved away from Gosport to outlying areas in a vain attempt to escape this maelstrom but in so doing sometimes the unwilling carriers of the disease themselves.

But the ancestry of the area in its present style really dates from the early 18th century. Gosport at the time described as having, 'two main streets with few pavements and overhanging buildings', one assumes in tudor style. Leading off these thoroughfares were numerous dark and narrow alleyways, the haunt of the press gangs and smugglers of brandy and tobacco. Meanwhile in 1746 the first bricks had been laid to the foundations of the Royal Naval Hospital at Haslar, to be completed several years later and reputedly the largest brick built building in the country at the time. Then around 1752 the navy established its main victualling and fuelling stores at Gosport so setting the final seal upon a town which has depended upon the senior service for its prosperity ever since. Accordingly, by 1777, the area is recorded as having some 5,000 inhabitants, a figure increasing many times in subsequent decades.

Neither was it all naval enterprise for the clay upon which much of the area stands was found suitable for brick making with vast brickfields and kilns set up at Wych Farm Bridgemary, Morelands, Leesland, Elson, Privett and Stokes Bay as well as perhaps the more obvious Clayhall. Sizable quantities of gravel were also excavated at Alverstoke, this locality being split between commercial exploitation and the wish to remain a quiet village favoured for the residences of retired naval officers.

But even though remnants of the earths natural industries may be seen today in the form of land depressions on the site of these workings, Gosport may rightly claim a place in history as being one of the precursors of the industrial revolution, even though little physical evidence of this remains in the town, for in Gosport Henry Gort patented a process in 1783 for the preparation of iron. This was so successful that iron production rose from a paltry 17,000 tons per year in 1740 to some 68,000 tons in 1788 and a staggering 250,000 tons by 1806. But this of course was not in Hampshire, as it had been found more economic to manufacture the backbone of Britain's industrial development in the midlands and north where reserves of the basic raw materials coal and iron ore were close at hand.

At the start of the 19th century there is evidence to suggest the population of Gosport had moved out of the immediate walled town to the suburbs, as the former had now given way to commerce.

Transport was available even if it was a time consuming affair, with stagecoaches running a regular service to both London and the West Country. To the capital for example one could board 'The Telegraph' from the offices of Mr. Padwick at the lower end of High Street (at the time known as Middle Street). This left at 5 am every morning and returned the following day. Another, 'The Yeoman', departed from the India Arms at 8.45 am, this also the starting point for the coaches travelling westwards. As is well known though the era of the stage-coach was to be short-lived, for elsewhere the first signs of a revolution in transport were being seen. The new 'railroads' had arrived, Gosport was destined to be one of the early destinations, receiving its railway in 1842, a few years after the commencement of the ferry and floating bridge service across the harbour to Portsmouth. Thus the stagecoaches were condemned to the history books.

The Guard Room of the former barracks at Forton and previously part of the old prison. *Author's collection*

Chapter Two

The Railway to Gosport

The story of the first railway to reach Gosport is today one which reads almost like a juvenile argument, the various parties involved seemingly oblivious to the harm they were causing to their own interests in what was a vain attempt to fight off an opposition.

To explain this fully it is first necessary to describe some of the earliest proposals for railway communication in the south of England. These began in 1834 with the passing of the London & Southampton Railway Companies Act for a line linking those two localities. This would travel over the land via Basingstoke, although a glance at a map will show how a route via Alton would have seemed to be a more direct and therefore seemingly logical approach. But the eyes of the L & SR were, even at this embryonic period, looking towards expansion. The Basingstoke route was chosen, allowing as it did, a hoped-for junction at that locality heading westwards towards Bristol. A similar branch from Barton (later Bishopstoke and then Eastleigh) to serve Portsmouth was also envisaged. But the proposed Bristol line was of course only a branch off the main route, hardly likely to encourage support from the merchants of Bristol who also had on offer the Great Western Railway scheme for a direct line from London. It was therefore left for parliament to decide and they came down in favour of the GWR proposals, although how much of this was due to the charismatic personality of their engineer, the redoubtable Isambard Kingdom Brunel, must be open to conjecture.

The L & SR had thus lost Bristol but still hoped to serve Portsmouth and attempted to cultivate support for their proposal in the naval port itself. The scheme was for a line to run via Fareham and Cosham, but not sufficient to satisfy the residents of Portsmouth who desired for themselves a direct line to London and certainly not a branch off another main route. It must be realised there existed at the time considerable animosity between the ports of Southampton and Portsmouth, the residents of the latter strongly against any railway company having the name Southampton in its title. There was also on offer a rival scheme supported by the newly formed London, Brighton and South Coast Railway Company. This was for a line from London through Dorking, Arundel and Chichester to Portsmouth, but again hardly a through route. The L & SR proposal went under the title of the 'Portsmouth Junction Railway Co.'. At the same time a branch was proposed from a triangular junction at Fareham to a terminus at Bury Road in Gosport. The Engineer to this was Francis Giles and the capital required for the main Portsmouth scheme was put at £300,000, the cost of the Gosport line was not recorded. Of this a certain amount was indeed promised upon subscription and numerous £3 deposits on shares paid. But the opposition by the people of Portsmouth proved too much to overcome and the main Bill failed in Parliament on 20th November 1838, taking with it the proposals for Gosport. The luckless speculators lost 12/6d of their original investment.

Shortly afterwards the rival LB & SCR scheme also floundered leaving Portsmouth with no projected railway at all. It was to be 1847 before rails were to reach the port and 1859 before direct communication was made via Guildford.

Despite the setback the L & SR were determined to press ahead with their expansion plans and so even before the main line was itself fully open to Southampton, attempted to reconcile matters by suggesting a name change for the company to that of the London and South Western Railway, at the same time projecting a railway from Barton through Fareham to Gosport. This was put to the shareholders at the general meeting of 23rd April 1839 when the main item on the agenda was 'Approving the draft copy of the proposed Bill for amending an Act relating to the Company and for making a branch line railway from Bishopstoke to Gosport'.

Why then may it be asked did they not try for Portsmouth again? The probable answer is because of the strength of opposition already encountered and the fact that at the time Gosport was seen as a reasonable alternative, passengers only having to make a short ferry journey across the harbour mouth to reach the trains – or so it was thought would be the case. Hardly an inconvenience when compared with the present day alternatives. There may also have been a degree of caution as to the likelihood of possible naval opposition where the Portsmouth railway would have passed through the 'Hilsea Lines', these were a series of man-made fortifications to the north of Portsmouth and at the time considered essential to the defence of the area.

To support the proposal the promoters had prepared a statement of existing and anticipated traffic to and from the area. This when read today provides a fascinating insight into the commercial life of the area at the time:

Passenger Transport

Between London and Gosport. 626 coaches per annum with an average compliment of 9 passengers per trip. Fare of 5/6d inside and 3/6d outside.

Between Gosport and Fareham. 626 coaches per annum with an average of 7 passengers per trip. The fare 1/- inside.

Between Southampton and Gosport. 1252 coaches per annum with an average of 7 persons per trip. Fare of 5/6 inside and 3/6d outside.

Goods

Portsmouth and Gosport to London. 5,000 baskets of mackeral, being equal to 250 tons at 1/6d per ton mile.

Portsmouth and Gosport to London. 10,000 lambs per annum at 2/6d per score.

London to Gosport. 750 oxen annually at 2/- each, destined for Gosport victualling yard.

Salisbury and the West Country to Gosport. 750 oxen per annum at 2/- each again destined for the victualling yard.

1,000 tons of goods were also recorded as passing by coastal vessel between London and Gosport at a rate of 14/- per ton. The journey taking 9-10 days.

In addition it was anticipated the following traffic would be attracted:

1. General goods traffic from London for consumption in the district into or through which they are brought.
2. Barley and malt, from Winchester and its neighbourhood.
3. Fish to the London market.
4. Lambs from the Isle of Wight.

5. Country killed meat to London.
6. Stores and provisions to Her Majesty's Victualling yard.
7. Garden produce.

Conspicuous for its omission is any mention of gravel or brick traffic emanating from the area.

Financial returns were anticipated as under;

	£ s d
Passenger revenue p.a.	26227 14 0
Parcels p.a.	1552 12 4
General goods p.a.	2929 12 0
Lambs p.a.	166 13 4
Mails (2 years)	584 00 0
	£31460 11 8

Accordingly, with opposition from Portsmouth the L & SWR dropped, a bill for a branch from Barton to Gosport received the Royal Assent on 4th June 1839, nearly a year before the last section of the main line was opened between Basingstoke and Winchester.

The Gosport branch involved raising £300,000 of additional capital, interestingly the same amount as had been claimed as necessary for the longer Portsmouth line. Even so the Directors had publicly stated that by building a line to Gosport they anticipated the cost to be only £180,000. Why then the additional £120,000? History fails to provide us with the answer. The new flotation was made up of 6,000 shares of £50 each, so providing a total share capital for the L & SWR of £1.7 million. Records have survived of the principle shareholders of the time and show that the majority of finance was attracted from either London or the north. This latter location in consequence of the profits then being made by the various mill owners. In contrast, the total stake of those having a vested local interest stood at only £262,500.

Construction of the new 15¾-mile line began almost at once, under the direction of Joseph Locke, then engineer of the railway company. He engaged Thomas Brassey as principle contractor who was well experienced in the railway field having been responsible for sections of the main London to Southampton line. Brassey brought with him a veritable army of 'navvys' (the name derived from the term navigators, meaning those who had built the canals) with little trouble envisaged through what was believed to be easy countryside.

So work progressed, with one major obstacle encountered in the form of a ridge in the earths surface extending east to west from Portsdown Hill to the Meon Valley and which the new railway would have to dissect on its course between Botley and Fareham. This was achieved by a tunnel 600 yards in length.

Opening day for the line was provisionally scheduled for 26th July 1841 but just 11 days before this, early in the morning of 11th July, a slip occurred within the tunnel over a distance of some 40 yards. (Locke and Brassey had confronted head on the peculiar strata over the area, being a type of clay which when dry required blasting to shift and yet when wet was little more than a river of liquid mud. To contain the substance, brick walls were erected, but in wet weather the slime just oozed over the top). Opening the line had to be postponed indefinitely. To compound their problems a further bad slip took place on the morning of 25th August when another 100 feet of tunnel caved in. Elsewhere the same subsoil caused embankments to be seriously affected.

Not surprisingly Locke was mortified. He reported to the L & SWR directors on 27th August 1841:

> 'I think you are already aware of the failure which took place a short time since in a portion of the Fareham Tunnel and of the delay in opening it produced. I beg to lay before you the following explanation, which may enable the proprietors to judge more correctly the extent of the failure and of the probable time for its reparation.
>
> The tunnel is upwards of 600 yards in length. It passed on the north side of the hill through beds of yellow and red clay, such as are usually found between the chalk and teriary formation and then enters the blue clay (similar to the London clay) through which it continues to the southern side of the hill. The failure occurred on the north side where the coloured beds of clay exist, just at a period when the last or junction length of earth was being removed for the brickwork and such was the weight of the slip and so immediate its effect, that notwithstanding every effort of the miners, the ground fell from the surface and carried with it a portion of the brickwork that had already been built. The thickness of the work was 3 feet.
>
> From subsequent examination, it appeared that the inverted arch and part of the side walls remained good – the arch requiring to be rebuilt for a distance of nearly 40 yards. The most vigorous measures for repairing the work were immediately adopted and these have been carried on night and day and will be continued until the whole is completed, which I expect will be accomplished in three weeks.'

Locke estimated the cost of remedial work to be £10-15,000. Both he and Brassey concluded the only feasible way to oblivate their difficulties was to abandon part of the tunnel, replacing it with a cutting. This work was undertaken, albiet hampered by reported, 'heavy and continuous rain for weeks on end.' In the process of which Brassey reported that he had suffered, '......financial ruin.'

In November 1841, Lt. Col. Sir Frederick Smith, on behalf of the Board of Trade, spent three days inspecting the new line. This was his second visit as he had previously attended the scene of the failure of 25th August. Sir Frederick reported his findings on 28th November concluding that; '......the ground has become almost semi-fluid with scarcely any slope of the cutting which would stand, while at the north end of the tunnel the slopes in consequence have lost all regularity of form and pour over the retaining walls upon the rails.' The remainder of the line did not attract adverse comment. Not surprisingly the conclusions reached by the Inspector were that the line should not be opened in its present state. But at the time the Board of Trade had no legal powers for enforcement of their recommendations and Locke mindful of the pressure being exerted on him by the directors to open the line and so obtain a return on their investment, agreed to bow to the pressures and opened the line to traffic the following day 29th November 1841.

There were some concessions though to the Board of Trades' recommendations. These included an overall speed limit of 20 mph with a lower limit on the unstable section of line. A continuous watch was also to be kept at Funtley, (north of the tunnel) where the same strata was causing similar difficulties. Trains would also run in daylight only with an inspection of the tunnel prior to the passage of each train. Little correspondence seems to have resulted from the railway inspectorate as a result of this decision, although today one may ponder upon the operating difficulties such restrictions were bound to present.

Unfortunately no records appear to have survived of the first public service to reach Gosport, although it is known that in addition to the ordinary trains 1,000 troops had been carried from Winchester to Gosport. But then only four days later disaster again struck with a slip occurring between the tunnels sufficient to bury the rails themselves. Locke visited the site of the failure the same day, 3rd December 1841, subsequently reporting his finds to the directors:

> 'On my arrival at the spot, I found the men busily employed in removing the earth so as to clear one line but, in consequence of the favourable weather and the risk there was in removing the earth so contiguous to the tunnel too rapidly, I thought it more prudent to order them to desist.......'

He continued;

> '........knowing the directors desire to make the safety of passengers paramount to all other considerations, will induce you at once to sanction the course I felt necessary to take this morning and for the present to close the line.'

Assuming the directors were indeed of such mind it may seem strange that they had allowed the railway to open in the first place.

But goods traffic continued to run whilst Locke and Brassey attempted to search for an answer to the difficulties. What was eventually arrived at is uncertain, although certainly these included sloping the cutting sides to a shallow angle which is still visible today. Then, on 26th January 1842, a letter was sent from the L & SWR to the Board of Trade stating their intention to re-open the line throughout on Monday 7th February 1842, enclosed with this was the statutory certificate from Locke that the line was, 'now ready'.

Re-inspection was undertaken for the Board of Trade by General Pasley on 4th February, covering not only the area around the tunnels but again the remainder of the line. (Consulted sources have stated Sir Frederick Smith was again responsible. Research undertaken with original source material contradicts this however. It is felt likely that Sir Frederick's signature appeared at the bottom of the report as senior inspecting officer.) As usual such a report is a valuable source of information including references to features not previously mentioned. One of these is a slip at Hawford (sic) where, '...... on the embankment a slip has taken place by which the 'up' line has been displaced for about 70 feet. This settled by supporting the rails on alternate courses of longitudinal and cross sleepers'. Apparently this was in similar fashion to Brunel's original 'baulk road'. The same report continues that Gosport station was finished although mile posts were still to be added. The previous speed limit of 20 mph to be retained,'....until the earthworks have become more consolidated.' One may wonder perhaps what state Gosport station was in at the time of the original opening in the previous November, or for that matter the preceding July when opening had first been proposed. Known sources fail to provide the answers.

The official Board of Trade conclusions were issued in a report dated 6th February and interesting by its attempt to take away the blame from the Inspectorate should a calamity occur later:

> 'I have not observed any cause for apprehension in respect of this tunnel in my present inspection of it, but in my former report I observed that the Lords of the Council (Board of Trade) could take no responsibility in such work as tunnels, even under ordinary and favourable circumstances, as their officials have no knowledge of the efficiency of the sectional strength of the sustaining arches and side walls and of the inverts, or of the quality of the work of which they are composed.
>
> The force of this observation applies much more strongly in such a case as the present where the work has already given evidence that it is exposed to great and sudden pressure and therefore, the whole responsibilities for using this tunnel must rest with the directors of the company, for they have the means, by frequent personal inspection and by reports of their engineer of arriving at a full knowledge of all the circumstances connected with this work and whether, if its form has undergone any change, that change has been to such an extent as to afford any grounds of apprehension for the safety of the work.'

The onus of responsibility was thus fairly and squarely returned to the railway company and more particularly Locke, there was little serious trouble at the site for some years to come.

The railway was thus finally opened, as intended, on Monday 7th February 1842, the first train arriving from Nine Elms, (then the London Terminus) with four first class carriages drawn by locomotive No. 17 *Queen*. It had taken 3½ hours to complete the journey, not at all bad when considering the operating conditions of the day. As was then customary, crowds thronged the station, the railway directors de-training and following a procession of 'band music' which wound its way through the streets and over the floating bridge to Portsmouth. The party then returned to the Star Hotel for dinner.

An engraving of the old Fareham station, probably around 1860, showing a train eminating from either Portsmouth or Gosport and the station building previous to re-building. *Dennis Tillman collection*

Another contemporary engraving this time of the Gosport terminus and, although containing several contentious features, is a good indication of the building that first existed. *Dennis Tillman collection*

So the railway had at last arrived, but at what cost? Certainly the delays caused by the slips at Fareham had taken their toll with the final bill for construction put at £404,271, some £220,000 more than once mooted. Of this total Gosport station itself was costed at £10,980, considerably more than the £1,391 allowed for Fareham. (Botley, the only other intermediate stopping place and located between Bishopstoke and Fareham was also opened at the same time although its cost is not recorded.)

The actual station building at Gosport had been erected by David Nicholson to a design produced by William (later Sir William) Tite. This is best described as comprising a stone and granite colonnade to contemporary Italian Tuscan origins, on one side of which a commodious train shed extended over both passenger and goods platforms. Tite had also been responsible for certain of the other original stations on the London to Southampton line, of these Micheldever (then known as Andover Road) is the best surviving example, although others notably Winchester and Southampton Terminus still stand. The design of Gosport was really a show of classic Victorian opulence, similar examples appearing at railway stations in various localities around the country.

It was now possible for travellers to venture from Gosport to London and return the same day. Four first, four mixed and two goods trains running daily in each direction. The fast trains conveyed 'through' London carriages and so avoided a change of train at Bishopstoke. At this point in railway history it was the custom to pay the appropriate fare relative to the speed of journey thus 22/- was the fare for a fast train; 21/- first class and 15/- second class on the mixed trains or if one had plenty of time and stamina there was an 8/6d fare for travel in the open wagons of a slow and stopping train.

Tickets as we know them did not then exist, instead intending passengers were required to enter their name and point of destination in a book. This was undertaken in a 'Book-in Office' from which the present term Booking Office was derived. A copy of the company regulations was also given to each passenger at the time of travelling. One of these regulations was that passengers had to arrive at the station at least five minutes before departure time and another that it was forbidden to open a carriage door or alight without the assistance of a member of staff. A veritable array of porters and policemen were employed to supervise the running of the trains.

In charge of these was Mr. Richard W. Stevens, his position akin to the present day stationmaster but at the time referred to as Station Superintendent. In addition there was Jethro Chandler in charge of goods traffic and John Callan in the Book-in Office.

G.A. Allcock in his book 'Gosport's Railway Era' of 1975, records the uniform and working conditions of the period and is a fascinating glance at the lot of the earliest railwaymen:

> 'Porters were dressed in fustian jackets (thick twilled short-napped cotton cloth usually dyed a dark colour) with arm badges and guards were resplendent in scarlet coats, lace collars and silver buttons. The railway policemen, who outnumbered the rest of the staff, wore swallow-tailed chocolate coats and tall hats with leather crowns. The remuneration of the ticket collector who worked two hours overtime on a Sunday was reported at 4½d.'

Obviously mindful of their previous comments, the Railway Inspectorate paid a repeat visit to the railway on 11th April 1842, this time the visiting officer, Col. Maberley found the line 'much improved'. Accordingly two days later, Mr. Martin, resident engineer to the line, was informed that the existing 20 mph speed limit could be dispensed with so allowing a minimum 41 minute journey time between Bishopstoke and Gosport. The Inspector's report concluding, '......the company being responsible for making the necessary arrangements for preventing accidents.......'

At the same time the line was assessed as to its suitability for the use of the 'passing of mails', approval for this seemingly required from the Board of Trade although why such a body was

DOWN BILL.

LONDON AND SOUTH WESTERN RAILWAY.

OPEN THROUGHOUT TO GOSPORT, PORTSMOUTH, & SOUTHAMPTON.

On and after 1st AUGUST, 1842,

THE TIMES OF DEPARTURE AND ARRIVAL ARE INTENDED TO BE AS FOLLOWS;

Down Trains. *Shewing the Time of Departure from NINE ELMS and other Stations, and intended Arrival at SOUTHAMPTON and GOSPORT.* — **On Sundays.**

	•† Mixed	• Mixed	Stopping	•† Fast Train	Goods	•† Mixed	Mixed	Fast	Mixed	Mixed	Mixed	Mail	Goods	Sundays: Mixed	• Mixed	Mixed	Mixed	Mixed	Mail.
	H. M.	H. M.	H. M.	H. M.	H. M.	H. M.	H. M.	H. M.	H. M.	H. M.	H. M.	H. M.	H. M.	H. M.	H. M.	H. M.	H. M.	H. M.	H. M.
NINE ELMS	7. 0	9. 0	10.15	11. 0	12. 0	1. 0	2.15	3. 0	4.15	5. 0	5.30	8.30	10. 0	9.15	10. 0	2.15	5. 0	7.15	8.30
Wandsworth	....	....	10.23	....	....	....	2.23	....	4.23	....	5.38	....	....	9.23	10. 8	2.23	5. 8	7.23	....
Wimbledon	....	....	10.32	....	....	....	2.32	....	4.33	....	5.47	....	....	9.32	10.17	2.32	5.17	7.32	....
Kingston	....	9.25	10.45	....	12.40	1.25	2.45	....	4.45	5.25	6. 0	8.55	....	9.45	10.30	2.45	5.30	7.45	8.55
Esher & Hamp. Co.	....	....	10.52	....	....	....	2.52	....	4.52	....	6. 7	9. 2	....	9.52	10.37	2.52	5.37	7.52	9. 2
Walton	....	....	11. 0	....	....	....	3. 0	....	5. 0	....	6.15	....	....	10. 0	10.45	3. 0	5.45	8. 0	....
Weybridge	....	....	11. 7	....	....	....	3. 7	....	5. 7	....	6.22	9.12	....	10. 7	10.52	3. 7	5.52	8. 7	9.12
Woking	7.54	9.55	11.25	11.46	1.25	1.55	3.35	3.46	5.25	5.55	6.40	9.27	11.59	10.25	11. 6	3.25	6. 6	8.25	9.27
Farnborough	8.18	10.20	....	12. 4	2. 5	2.20	....	4. 4	....	6.20	....	9.50	12.41	....	11.35	....	6.35	....	9.56
Winchfield	8.30	10.35	....	12.17	2.25	2.35	....	4.17	....	6.35	....	10. 8	1.13	....	11.52	....	6.52	....	10. 8
Basingstoke	8.50	11. 0	....	12.33	3.10	3. 0	....	4.33	....	7. 0	....	10.31	1.53	....	12.15	....	7.15	....	10.21
Andover Road	9.25	11.30	....	1. 1	4. 5	3.30	....	5. 1	....	7.30	....	10.59	2.39	....	12.47	....	7.47	....	10.59
Winchester	9.43	11.50	....	1.19	4.40	3.50	....	5.19	....	7.50	....	11.16	3.14	....	1. 7	....	8. 7	....	11.16
Bishopstoke	10. 0	12.12	....	1.37	5.20	4.12	....	5.37	....	8.12	....	11.31	3.35	....	1.27	....	8.27	....	11.34
SOUTHAMPTON	10.20	12.30	....	2. 0	6.10	4.30	....	6. 0	....	8.30	....	11.57	4. 0	....	1.45	....	8.45	....	11.57
BISHOPSTOKE	10. 0	12.12	....	1.37	5.37	4.12	....	5.37	....	8.12	....	11.44	3.35	....	1.27	....	8.27	....	11.44
Botley	10.13	12.24	....	....	....	4.24	....	....	....	8.24	....	11.56	....	....	1.39	....	8.39	....	11.56
Fareham	10.25	12.38	....	1.58	....	4.38	....	5.58	....	8.38	....	12.10	....	....	1.53	....	8.53	....	12.10
GOSPORT	10.45	1. 0	....	2.25	6.25	5. 0	....	6.25	....	9. 0	....	12.32	5. 0	....	2.10	....	9.10	....	12.32

FARES.

Passengers. — **Horses & Carriages.**

STATIONS.	Distance.	FAST TRAIN. 1st Class.	FAST TRAIN. Liv. Servts	MIXED TRAIN. 1st Class.	MIXED TRAIN. 2nd Class.	THIRD CLASS.	CARRIAGE.	1 HORSE.	2 HORSES.	3 HORSES.
	Miles.	s. d.	s. d.	s. d.	s. d.	s. d.	s. d.	s. d.	s. d.	s. d.
London to Wandsworth	3	..	..	1 0	0 6	..		not	taken	
.... Wimbledon	6	..	..	1 6	1 9	..		not	taken	
.... Kingston	10	..	..	2 0	1 6	..	10 0	7 0	10 0	12 0
.... Esher & Hampton Court	13	..	..	2 6	1 9	..	10 0	7 0	10 0	12 0
.... Walton	15½	..	..	3 0	2 0	..		not	taken	
.... Weybridge	17½	..	..	3 6	2 6	..	10 0	7 0	10 0	12 0
.... Woking	23	6 0	5 0	5 6	4 0	2 6	12 0	8 0	12 0	15 0
.... Farnborough	31½	9 0	6 6	8 0	5 6	3 6	17 0	11 0	17 0	21 0
.... Winchfield	38	10 6	8 0	10 0	7 0	4 0	21 0	14 0	21 0	26 0
.... Basingstoke	46	12 6	9 0	12 0	8 0	4 6	26 0	18 0	25 0	30 0
.... Andover Road	56	15 6	11 0	15 0	10 0	5 6	31 0	22 0	30 0	36 0
.... Winchester	64	18 6	13 0	17 6	12 0	6 6	36 0	25 0	34 0	42 0
.... Bishopstoke	72	19 6	14 0	18 6	13 0	7 6	39 0	28 0	37 0	46 0
.... Southampton	77	21 0	15 0	20 0	14 0	8 0	42 0	30 0	40 0	50 0
London to Botley	77	20 6	15 0	19 6	14 0	8 0	42 0	30 0	40 0	50 0
.... Fareham	82	21 6	15 6	20 6	14 6	8 6	43 0	31 0	41 0	51 0
.... Gosport	87	22 0	16 0	21 0	15 0	8 6	44 0	32 0	42 0	52 0

Children under 12 Months old no charge: above 1 Year and under 10, half price.

From SOUTHAMPTON to GOSPORT,

Shewing the Time of Departure from SOUTHAMPTON and other Stations, and Arrival at GOSPORT.

	Mixed	Mixed	Mixed	Mixed	SUNDAYS. Mixed	SUNDAYS. Mixed
	H. M.	H. M.	H. M.	H. M.	H. M.	H. M.
From Southampton	9 . 45	12 . 0	4 . 0	8 . 0	1 . 10	8 . 10
Bishopstoke	10 . 0	12 . 12	4 . 12	8 . 12	1 . 27	8 . 27
Botley	10 . 13	12 . 24	4 . 24	8 . 24	1 . 39	8 . 39
Fareham	10 . 25	12 . 38	4 . 38	8 . 38	1 . 53	8 . 53
Gosport	10 . 45	1 . 0	5 . 0	9 . 0	2 . 10	9 . 10

FARES.

From Southampton to Gosport.	Distance	MIXED TRAIN 1st Class	MIXED TRAIN 2nd Class	CARRIAGE	1 HORSE	2 HORSES	3 HORSES
	Miles.	s. d.	s. d.	s. d.	s. d.	s. d.	s. d.
From SOUTHAMPTON to Bishopstoke	5	1 6	1 0	10 0	7 0	10 0	12 0
Botley	10	2 6	2 0	10 0	7 0	10 0	12 0
Fareham	15	3 6	2 6	10 0	7 0	10 0	12 9
GOSPORT	20	4 6	3 0	10 0	7 9	10 0	12 0

Children under 12 Months old no charge; above 1 Year and under 10, half-price.

LONDON TIME observed; and the Doors of the Stations finally Closed, Five Minutes before the time of Starting.

N.B. Passengers with any large quantity of Luggage should be at the Station at least fifteen minutes before the time of departure. Private Carriages and Horses (a day's notice being given) must be at the Station twenty minutes before the time of departure of the Train. Omnibuses and Stage Coaches, conveying Passengers and their Luggage, must be at the Station ten minutes before the time of starting, so as to allow time for the removal of the Luggage from such Omnibuses or Stage Coaches on to the Railway Carriages, and thus prevent delay in the time of the Train starting.

The *First Class Trains* convey *First Class Passengers only*, excepting that accommodation is afforded for a limited number of Servants in Livery. These Trains will not call at any Station between London and Woking Common, but will take up and set down Passengers at all the Stations West of Woking Common.

The *Mail Trains* call at Kingston, Esher, Weybridge, Woking Common, and all Stations beyond Woking Common. The Fares the same as the Mixed Trains for both First and Second Class Passengers.

The *Mixed Trains* will stop at Kingston only, between Nine Elms and Woking unless *by Signal* for Passengers going to the west of Woking Common Passengers by *mixed* Trains riding in their own carriages, will be charged Second class fare only.

The Short Trains stop at all the Stations.

Third Class Passengers will be taken by the First Train every Morning, except Sundays.

Passengers are allowed to carry, free of charge, a certain quantity of luggage, which is understood to be **personal baggage**, and not **merchandize.**

First Class 112 lbs.
Second „ 56 „
Third „ 28 „
all excess 3s. per Cwt.

involved in what appears to be a commercial transaction is unclear. Apparently all was satisfactory for sanction was duly given. However, less than two months after this, in June 1842 a train arriving at Gosport was unable to stop and collided with the far wall demolishing part of it along with some adjacent gates. The report into the accident concluded that the guard was primarily responsible, this unfortunate individual was fined 40/- and severely reprimanded. It added that his punishment had been lenient due to 'his van brake working in a contrary direction.' The original cause of the trains late appearance was attributable to engine failure so it may have been the case that the crew were attempting to make up time.

A different kind of justice was handed out to two juveniles appearing in front of local magistrates at Gosport Police Court later the same year. The children concerned having been caught throwing stones at passing trains. For the offence the two received three months and one month custodial sentences.

By the summer of 1842, surviving timetables, or 'bills' as they were then called, show nine trains daily working in each direction. Of these, two being goods services and supposedly running non-stop between Gosport and Bishopstoke in either direction. This though must be doubtful for the journey time is shown as occupying some 1¼ hours in either direction compared with the 41 minutes for other services inclusive of two intermediate stops. The timetable is interesting for it shows other aspects as well; details of the Sunday service and the fact that horses and carriages (horse drawn road type) could be carried at this early stage in the lines history for instance.

As already recorded, the Gosport line had been in the forefront of various advances in technology. An example of this being early in 1844 when at the half yearly meeting of shareholders, approval was given for the installation of the electric telegraph throughout between Nine Elms and Gosport. This, however, was not primarily for signalling or for regulating trains but instead for the benefit of the naval authorities who wished to supersede the then method of communication between the Portsmouth area and London which was then the semaphore. The visual impression of men waving flags at each other from vantage points across the country is somewhat amusing today! Accordingly a contract was entered into with the foremost manufacturers and installers of such equipment, Messrs. Cooke and Wheatstone, they having already supplied similar items to other railway companies. The telegraph was brought into use on 1st April 1845, and became the longest communications link of its type in Britain. A group of clerks were employed at the stations to send and read messages. The public too were allowed access to the benefits of this advancement in communications; on payment of 3/- or 5/- for a single or 'replied paid' message having a maximum length of 40 words. Delivery was extra. It should be added that the telegraph was itself an instrument by which the passing of an electric current flicked a needle in one direction or another. It was possible to code the various letters and numbers by means of a simple system. Examples of various early telegraph instruments can be seen on display at the National Railway Museum, York.

The history of Gosport itself during the nineteenth century is interwoven around the story of royalty visting and coming through the area. This is believed to have started on 8th October 1844 when Prince Albert travelled by train to Gosport to greet Louis Philippe of France who had arrived at the Victualling Yard on a state visit to England. The royal party, accompanied by various dignitaries and other VIP's, then travelled to Farnborough by rail before completing their journey to Windsor by carriage where the Queen was waiting. It would appear a favourable account of the arrangements concerning the railway company was reported, for upon Louis Philippe's return to France, later in the year, Queen Victoria accompanied her guest by rail to Gosport station.

G.A. Allcock records the occasion as having been a terrible day, the usual bunting displayed totally ruined by the downpour and associated high winds.

Louis Philippe's departure from this country was to have been by boat from the Victualling Yard, but it was decided to delay the journey temporarily to allow the conditions to improve. Whilst this occurred the Superintendent of the Victualling Yard, Thomas Grant, was requested to supply refreshments to the Royal party. Unfortunately nature continued to do her worst and it was eventually decided to alter the travel arrangements by returning to London and then attempting the journey via Dover, this proved more successful. Grant was later rewarded with a valuable piece of plate and later a Knighthood.

It was shortly after this, in 1845, that Queen Victoria purchased the Osborne Estate near Cowes on the Isle of Wight. The Royal family were regular visitors to their new acquisition from then on until the time of the Queen's death. Published sources would seem to imply that the Victualling Yard pier was the favoured embarkation point for these journeys although why Portsmouth and Southampton were not used is not completely clear.

Contemporary records imply that at the suggestion of the Royal Consort, Prince Albert, the L & SWR agreed to a short

A telegraph instrument of the early days. The two handles were able to indicate the various letters of the alphabet according to a variety of codes.
British Rail

extension of railway from the existing Gosport station to a point nearer to the actual pier. This some 605 yards long and opened for traffic on 13th September 1845. The work was inspected by Maj. Gen. C.W. Pasley for the Board of Trade a few days later, his report confirming the lines intended use as well as what would have also seemed likely for the carriage of goods to the naval yard;

> 'Extension into Clarence Victualling Yard, thereby obtaining a communication with Portsmouth Harbour for the use of Her Majesty and for Government purposes exclusively.
>
> This extension is 605 yards in length proceeding from the terminal station by three curves of 33,26, and 16 chains radius respectively, all bending to the northward and by three descending gradients of 1 in 112, 1 in 164 and 1 in 133 respectively. The first crosses the turnpike road from Gosport to Fareham, on a level and then cuts into the glacis and passes the ditch and ramparts of Gosport lines by a wooden viaduct bridge and an archway, beyond which it crosses the road from Gosport to Clarence Victualling Yard on a level and then passing over a field, the property of the Ordnance, it meets the harbour, a small part of which it passes by an embankment of about 50 yards in length and 12 ft. high, the foot of which is dry at low water and finally enters the Victualling Yard, where it terminates near the jetty. With the exceptions of the crossings of the ditch and rampart and the embankment above mentioned, the rails run nearly along the level of the original ground. The viaduct for the passage of the ditch consists of 10 bays of about 21 ft. each in clear width, supported by piers of three piles, where it crosses the wet part of the ditch and of three posts on a sill and dwarf wall for the dry part of it. Each pier has a cap of whole timber over the heads of the piles or posts, supporting longitudinal bearers, four in number, resembling those of the Bricklayers Arms viaduct of the South Eastern Railway, two of which are of whole timber immediately under the rails, while the two others are of half timber on each side. The superstructure consists of 3 inch planks, the ends of which have longitudinal wales of timber over them about 12 inches high, instead of parapet railings. The gateway under the rampart is an elliptical arch, 15 ft wide, built of bricks and cement, in the same curtain, and a little to the northward of one of the present public roads and gateways leading from Fareham to Gosport. The viaduct and gateway are both of ample strength; and this extension, which consists only of a single line of rails, is in a safe and efficient state. There is a short siding near the jetty for the engine and tender to run into, whilst the carriages pass on. A reception road for the accommodation of Her Majesty is proposed to be built at the extremity of the extension, which will render it much more convenient. Gates shut across the rails, on each side of the two level crossings before mentioned, but not across either of the roads, which would be superfluous!

Within the yard was a single curved platform 520 feet long, covered throughout its length and flanked on one side by a long wall. A waiting room provided accommodation for the Royal travellers. The station was referred to as 'Royal Victoria Station' and retained for the exclusive use of monarchy. The waiting room was located at one end of the station, from which it was a short walk at right angles along a covered gangway before a further right angle turn took the party down a gangway and on to the pontoon. The waiting rooms were themselves furnished with stores taken from the Clarence Yard itself, whilst on each occasion a royal journey was made, a carpet was also provided to cover the distance from the Royal coach to the Royal Yacht. It is doubtful that much use was actually made of the waiting rooms themselves. The occasion of the first use of the new facilities is unfortunately not recorded, although it is known that long periods were spent at Osborne during July and August and December to February each year. As was associated with the period a considerable amount of planning went into the arrangements for a Royal Train journey, details of similar arrangements are given on page 20.

This then had been the first stage in the development of the railways of Gosport, a saga that would eventually involve the construction of two completely new lines within the town's boundaries. Part of the reason for this was of course the success the railways had generated themselves. Coal for example was now cheaper to purchase as a result of reduced transportation costs; there were no less than ten coal merchants operating in Gosport by 1852. Another reason was the facilities that were now available to the masses for fast and relatively inexpensive travel compared with the stagecoaches. The latter though were not completely dead, for the coach owners continued to attempt to compete, but it was a battle already lost. Some then switched their attentions to the short-haul business. An omnibus met each train and took passengers the short distance to either the High Street or the Pier for the cross harbour ferry, with a similar arrangement for the return.

The line to Clarence Yard shown passing under the old town walls. It was these fortifications which caused the railway to stop short at its existing terminus in 1841. *D. Cullum*

Above: The Royal station at Clarence Yard, probably around 1900.
British Rail

Right: A close-up of the Royal waiting room, reputedly never used by Queen Victoria. *Gosport Museum*

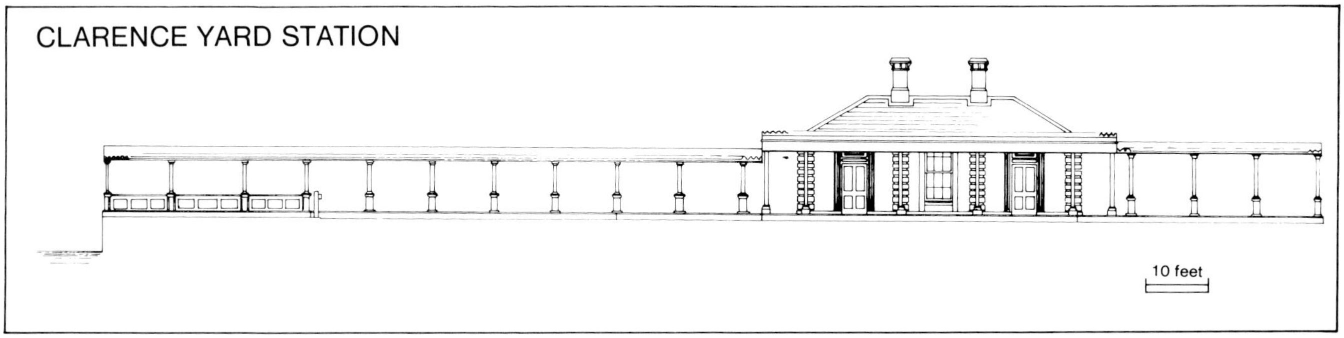

Chapter Three

The Line to Stokes Bay

The harbour ferry and town of Portsmouth as seen from the Gosport side. Prior to the Stokes Bay railway and pier, Portsmouth was the nearest means of embarkation for the Isle of Wight. *Author's collection*

With the Royal Family having shown approval of the Isle of Wight as a holiday venue, the railway companies recognised it as likely to be also a favourite venue for the masses as well. Consequently, as early as 30th October 1846, the L & SWR Board gave approval for a survey to be undertaken for a proposed 'Gosport Pier and Branch Railway', this was to be about one mile in length and terminating in a timber pier some 300 yards along. Cost was estimated to be at 'not more £15,000'. A little later a comment appeared that the Board would go ahead subject to the approval of a J. Coche. Possibly this individual was not in favour, for nothing else was heard of the scheme. Part of the reason for this may have been that at the time the railway company were involved in one of their seemingly endless battles with the rival GWR, certainly not in the Hampshire area, but it would appear that all available energies were directed towards this rather than an extension of their system to fill what was a definite need.

At the time the only means available to rail passengers to reach the Isle of Wight was by train to Southampton and thence by steamer across the Solent. In 1847, however, the LB & SCR opened their line from Havant to Portsmouth, followed the next year by the South Western line from Fareham to Portsmouth. The island became more readily accessible even if a tempestuous passage across Portsmouth was a feature. On this point the reminiscences of a traveller referring to such a trip in 1872 are of interest, for undoubtably the same criteria would have applied a few decades before;

> '........the journey (from London to Ventnor) was rarely accomplished under five to six hours. The scramble at Portsmouth and Ryde for tram accommodation was not an agreeable experience and with a party of children, nurses with luggage, perambulators and bath tubs, a man had to be something of an organiser to get through without losing his temper and some of his belongings as well.'

Into this vacuum came two entrepreneurs, William Pearson and Charles Ridout who on 21st November 1854 registered their intentions for a line to Stokes Bay from Gosport under a grandious and long-winded title. The shortened version of the name more commonly used is 'The Stokes Bay and Isle of Wight Railway and Pier Company.'

In essence the new Company's plan was similar to the proposals of eight years before and involved the construction of 1½ miles of railway from the main line just west of Gosport station to a new pier on the coast at Stokes Bay, trains to run directly onto the pier where ease of transference of passengers and luggage could take place. From that point it was only 2½ miles to Ryde, where a new pier was to be erected. Support was, it would seem, to be relatively forthcoming and by January 1855 the route had been surveyed and estimated to cost some £30,000. Landowners affected had also been approached and in the main gave their support. It seemed at this early stage as if everything was going to go smoothly.

But perhaps as an indication as to future fortunes the Company's bankers, Messrs. Paul Strahan & Co., failed shortly afterwards involving the loss of just over £100. Furthermore a considerable amount depended upon the attitude of the L & SWR, which indicated that certainly no financial support was possible. In the meanwhile the engineer, Hamilton Henry Fulton, had arranged for a firm of contractors, Messrs. Smith and Knight, to agree to build the line at a fixed price. The contractor in addition agreeing to subscribe some £14,000 of capital, provided that the majority of the remaining capital was taken up by other subscribers. This may today seem strange but at the time was a well recognised method of conducting business even if in this particular case the relevant percentages were somewhat unusual.

Not unexpectedly considerable opposition to the proposals was received from the Portsmouth area as well as the ferry company then operating between that locality and the Island, although a deal was later struck whereby such opposition would be dropped if the new company would concede the building and operation of their own ferries and in addition allow others to use the new facilities. To this the Stokes Bay directors readily agreed, for besides quietening the opposition it also reduced their capital requirements by a fifth to only £24,000.

Despite a degree of concern expressed by Smith & Knight over how the remainder of the capital could be raised, the company pressed ahead with their objectives, the Stokes Bay Railway and Pier Act receiving the Royal Assent on 14th August 1855. Expenses were put at a total of £20,000 of which £7,000 was estimated for the cost of Stokes Bay Pier itself. The share capital was set at £24,000 which should then have provided more than enough finance. The work was to be completed within three years.

Somewhat suitably, the first General Meeting of the new company was held in the Anglesey Hotel, Alverstoke on 3rd October 1855, very close to the route of the proposed railway. The Chairman, Thomas Fleming of Southampton, reported that;

> '.......there was every probability of it (the railway) being used by Her Majesty in preference to the present circuitous and unpleasant route from Gosport.'

He continued:

> 'Prince Albert had visited the site of the pier, in company with General Sir F. Smith, of the Royal Engineers.'

Little though is known of this visit.

The Stokes Bay company were in the meanwhile trying to get Smith & Knight to commence construction, the contractors having raised their estimate of cost to £16,200 for the line with an additional £8,000 for the pier. Not unexpectedly the engineer was requested to try and find a different contractor who would be prepared to build the line for a less figure and/or take a larger proportion of the cost in shares but this proved to be a far harder task than had first been envisaged.

But it was not all bad news for on 21st July 1856 the 'Isle of Wight Ferry Company' had obtained powers to build a chain ferry from the proposed Stokes Bay pier to Ryde, such a venture could only benefit the railway, although one is left to ponder somewhat at the idea of chains across what was even then a very busy shipping lane.

Still work on the railway had not begun, the latest estimate of costs now put by the engineer at £40,000, the increase attributable to changes dictated by the L & SWR to ensure safe passage of traffic. The land required costed at some £10,735. Smith & Knight had by now decided to completely abandon their involvement in the railway company no doubt in some way relieved because of the amount of aggravation that had been caused. Time though was running out and it was necessary to arrange for Acts authorising an extension of time in 1858 and again the following year. At the same time various minor changes were made to the authorised route, no doubt in an effort to reduce costs.

From contemporary records it would appear the railway company had never had the intention to own their own rolling stock and so operate their own services. Instead, trusting that agreement could be reached with the South Western. This came one stage nearer on 9th March 1858 with an agreement of four clauses arranging for the L & SWR to work the line for a fixed rental of £1,600 p.a., the figure later increased to £1,800. No mention is made though as to number of trains to be provided and of course there was at this time still no railway to operate.

At the same time two new tenders were received for construction, one from a Mr. Catlin for £34,000 and the other for £28,600 from Messrs. Lucas. The lower tender was accepted, the company having been authorised to increase its share capital by some £16,000 as a result of the 1859 Act, which therefore covered the increased cost. As was practice there were also borrowing powers equivalent to one third of the total share capital, to be taken up as soon as half the original shares had been subscribed. But the Stokes Bay directors and Lucas were unable to agree full terms and it would appear he simply faded from the scene. Shortly afterwards a tender was received from Brassey & Ogilvie, this an offshoot from the company responsible for building the original Gosport line. They somewhat surprisingly agreed to accept shares as full payment for their work. It would appear the Stokes Bay Company hardly considered a venture worthy of much support from other investors.

Construction thus began at last, although records of it are rather sketchy. Certainly with easy terrain covering the two-mile route little difficulty was encountered, the main engineering features were the Workhouse and Little Anglesea viaducts and of course the pier itself.

In the midst of this activity the L & SWR and LB & SCR had entered into a new agreement with regard to the pooling of traffic and rates charged and this was to have a profound impact upon the fortunes of the Stokes Bay line in later years, although details of the arrangements were not at the time made public – more is written of this on page 20.

With the new line almost completed, the customary notice of intention was given to the Board of Trade. This resulted in Captain Tyler visiting the route on 5th January 1859. Unfortunately a full inspection was not possible, for in a dispute between the various parties the contractors had removed a rail. The preliminary report then read as follows:

> 'I have the honour to report........I yesterday inspected the Stokes Bay Railway.
>
> This is a double line, a mile and a half long, extending from a Junction with the London and South Western Railway near Gosport to a Pier which has been constructed in Stokes Bay. The steepest gradient is 1 in 96½, and the sharpest curve has a radius of 12 chains. The permanent way is laid with double-headed rails weighing 75lbs to the lineal yard and 21 ft. long. The points of the rails are fished and the chairs which support the intermediate positions are secured to transome sleepers by wrought iron spikes. There are several viaducts constructed with cast iron girders upon timber supports – I was unable to test these or the pier in Stokes Bay, because the agent for the contractors declined to allow an engine to come upon the line. I observe that a raised stage is

required at the junction with the South Western line near Gosport; that there are certain level crossings of what appear to me to be public roads which have not been authorised by Parliament; and that the terminal station at the pier is not quite completed.

I have therefore to report my opinion that the line cannot be opened by reason of the incompleteness of the works, without danger to the public using it.'

Shortly afterwards the Isle of Wight Ferry Company attempted to commence services between Ryde and Stokes Bay, a ferry trying to berth alongside the new facilities but prevented from doing so by a number of navvies, a similar obstruction met the railway directors who had made their way from the Anglesea Hotel to meet the group.

Evidently the problems between the parties were settled, for a re-inspection took place on 30th March 1863. Captain Tyler again visited the line commenting that providing strengthening took place to the girders on the pier he would sanction opening of the line to traffic. The inspection was followed by a reception unusually provided by the contractors, although they did this time invite the railway directors.

various exchanges took place with the South Western Board. As expected there was little improvement, the larger company well aware of its obligations to its neighbour, the LB & SCR at nearby Portsmouth and certainly had no wish to evoke conflict with that company as well. The easy way out was to do little and that was in fact the case although an amusing sideline comes from the traffic manager of the South Western who stated he was;

'......disgusted with the Stokes Bay Company......really did not know who comprised that Company, for every letter he received was from a different person.'

But that did not stop a deputation of directors of the Stokes Bay Company approaching the South Western Traffic Committee on 27th October 1864. Their complaints covering three separate issues:

1. Increased pier accommodation.
2. Faster trains between Stokes Bay and London as well as from Stokes Bay to Havant. (This latter request would seem somewhat unreasonable from the outset.)
3. Completion of the new pier at Ryde.

A poor quality but very rare view of a train on Stokes Bay pier around 1895. Just discernible are the various buildings on the actual pier with a ferry on the extreme left. *Bert Moody collection*

One week later, on Monday 6th April, the line opened to traffic without, it is reported, any kind of ceremony. Instead the first train left Waterloo at 8 am, and arrived at Stokes Bay at 10.55 am, the passengers then transferring to the steamer, *Garloch*, for a 15 minute crossing to Ryde. This timing included a reversal at Gosport itself as the Stokes Bay line was then only accessible from a trailing connection into the main line.

Public services began with five trains each way on weekdays and two on Sundays, the trains operated by the engine the South Western had stationed at Gosport. Coinciding with this were five ferry sailings weekdays with three on Sundays, whilst in addition there was for a matter of weeks a service between Stokes Bay and Cowes. All though was not well, for despite having the advantage of a direct train-to-ferry exchange, passengers were at first required to change trains at Bishopstoke. The journey thus showed little advantage compared with the existing route through Portsmouth.

Not surprisingly the Stokes Bay directors were not entirely happy with the service provided and in an attempt to redress this, In reply Mr. Scott promised to consider the requests, any decision was later postponed indefinately.

The arrival upon the scene of the new route and associated ferry service had caused some consternation in Portsmouth itself. The various interested parties having watched from the sidelines as the scenes were enacted, there appearing to be little likelihood of the railway ever being built. It was now time to act and no time was wasted in obtaining powers for what was to be the first tramway in Portsmouth, authorised in June 1863, although not operating until May 1865. (More of the story of the rival Landport and Southsea tramways and subsequent Southsea Railway are told in the 'Southsea Railway' by the present author and published by Kingfisher Railway Productions.) The Joint Committee (comprising representatives of both L & SWR and LB & SCR railways, set up to administer the lines in Portsmouth), were also alert to the threat from the new line, seeing a fall in Isle of Wight traffic through Portsmouth as a distinct possibility. The culmination of this was an improvement in services and eventual building of the harbour station, although this was not opened until 1876.

Stokes Bay pier from the foreshore and clearly showing the landing area. The building at right angles to the train shed was a later addition by the Admiralty after train services ceased. *Sean Bolan*

But all this was some years in the future and it would seem that now should have been the time for the little company to consolidate its position so as to be ready to take on the opposition of future years. This, however, did not occur, partly it would appear because of their lack of management ability, well illustrated by the comment made by the South Western traffic manager and partly by difficulties involving the ferry company albeit out of the control of the hands of the railway directors.

It is worth mentioning briefly that the ferry company were, by June 1863, using three steamers for the crossing. *Garloch* has already been mentioned, but there was also *Chancellor* and *Victoria*. The last two had disappeared by August 1863 at which time the original vessel was in need of repairs and hurried arrangements had to be made to borrow a suitable replacement from the rival Portsmouth operators. The IWFC attempted to solve the situation by raising additional capital, after all their new pier at Ryde was still incomplete. But potential investors were put off by the thought of such a small concern becoming dangerously over-subscribed, little additional finance was therefore forthcoming.

By early 1865 the service was being operated by a steam tug, hardly suitable for the purpose intended and the object of ridicule, especially in Portsmouth where the unfortunate *Ursa Major* was referred to as '...the tub...'. Luckily this did not last long and early in 1866 it was reported that a 75-ton paddle steamer, *Her Majesty* had taken over.

An Edwardian group at Stokes Bay in 1911 with the railway embankment in the background. *Author's collection*

The next stage in the story comes with formation in April 1864 of a new company whose intention was to take over both the Stokes Bay railway and ferry companies and in so doing attempt to complete the unfinished works at Ryde. Little else is heard of the concern who were finally dissolved in 1882. But before this the situation of the IWFC had become desperate, an enquiry at Ryde finding them unable to fulfill their obligations with the inevitable result that the company was wound up in May 1865. As a postscript to the ferry company one can do no better than quote from the Ryde town guide for 1909, when referring to the remains of the incomplete pier. It described as having:

> 'Graduated enclosures for ladies and gentlemen...... and non swimmers could enjoy the novelty of disporting themselves a quarter of a mile out at sea without the risk of being drowned.'

Neither was it only the ferry service that was causing difficulties, for the problems of a reversal at Gosport and the lack of suitable accommodation at Stokes Bay were making it difficult to deal with the existing traffic. Additionally there were problems with the steamers being unable to tie up at the pier during inclement weather. Both railway companies agreed that change was necessary although the amount of financial responsibility to each party could not be agreed upon.

After the usual round of accusations as to whose fault the present situation was, both parties set about on what was to prove to be the start of the final round of negotiations. The South Western offered in 1871 to buy out the Stokes Bay Company in exchange for £35,000 of ordinary South Western stock. The little company responded by demanding a better deal, £40,000 in addition to an increase of £200 on the annual rental of £1,800 already paid, this latter figure itself unchanged from 1858.

Surprisingly there was hardly any opposition from the South Western and so approval of the shareholders to a sellout was apparently readily obtained. They recognising, that despite a gradual increase in traffic, that there was little chance of the line providing much scope for future expansion. Accordingly it was expected that the transfer of ownership would progress rapidly, although delays in changing the lease of some war department land at Stokes Bay dragged on for a further three years and it was not until 11th June 1875 that final completion took place. The Stokes Bay Railway Company having had a life of just under 20 years.

The purchase was reported to the South Western's own shareholders in the following fashion:

> 'Looking at the importance of the Stokes Bay Railway, as affording the shortest and most convenient communication between the mainland and the Isle of Wight, the directors do not hesitate to recommend the proprietors to sanction these arrangements.'

Strange sentiments indeed when compared with the contempt previously expressed against the line. One may ask then, was there an ulterior motive in such a move, perhaps seeing the possibility in retaining a separate route for the lucrative Isle of Wight traffic should the joint Portsmouth agreement become tarnished. History fails to recount the answer.

What then of the Stokes Bay shareholders? Their's was the fate of so many in similar situations elsewhere. The distribution of funds on a scale varying from 16/- in the £ on mortgage bonds to just 4/- on ordinary shares. Total expenditure on the line since 1855 having been something over £76,000 against receipts of just £21,724 accrued from the time of opening.

In retrospect it was the only decision possible, the lines former owners now being able to sit back and watch the meagre gestures made to improving the railways fortunes until they were suddenly cut short a few years into the twentieth century. Had they waited until then for a sellout, there would have been no cake left for distribution at all.

Chapter Four

Decades of Expansion

On the main line the passenger services had, by 1859, been increased to no less than nine trains daily in the 'down' direction although only six corresponding workings are shown in the 'up' direction. The service was advertised as running from Southampton to Portsmouth via Bishopstoke. With certain exceptions passengers had to change trains at Fareham for the local Gosport service.

What was then a decline in passenger services compared with that of a few years before was hardly likely to go un-noticed, several letters of complaint being received by the L & SWR concerning the 'transfer of trains' at Fareham and one written by the Rev. J. Walpole over the '......quality of carriages used on the Gosport line......'. In an effort to placate these criticisms, through carriages were reinstated from London, although correspondence on the same subject was to continue for many years to come. One may reasonably assume that by the time the Stokes Bay line opened matters had improved somewhat, although now the complaint was over the reversal for Stokes Bay trains at Gosport itself.

Accordingly, only 18 months after the new line opened, the South Western directors acceded to pressure and recommended a spur be provided so as to form a triangle at Gosport and allow Stokes Bay trains to proceed direct to the pier without reversal. At the same time sanction was given for a 'small station' on the Stokes Bay line, the purpose of this to accommodate Gosport passengers who might otherwise find themselves deposited some two miles away from their intended destination. Around the same time further changes had been approved with sanction given for ticket platforms, a new facility at Stokes Bay, although the existing item at Gosport was to be lengthened. Ticket platforms themselves were locations where a brief stop was made just prior to the main station. They were not normally for the use of the travelling public and instead a number of ticket collectors would board the train during its brief halt to examine tickets prior to arrival at the main station. The objective was that passengers were afforded uninhibited passage from the train upon its eventual arrival. A modification to the scheme is shown in the 'open station' policy of some areas today.

The directors had also sanctioned on 8th June 1863 a new station between Gosport and Fareham. This was stated to be '......for the accommodation of troops......' occupying the various forts at the north end of the town. Approval was given for a stopping place with two short platforms, Brockhurst station opening on 1st November 1865 with the existing gateman to act as stationmaster. Two local trains were provided for the initial service. Interestingly, on 17th March 1864, a request had been received by the South Western from the residents of the Anglesea area for Stokes Bay trains to stop at a platform to be erected by

A superb study of Brockhurst prior to the advent of the Lee-on-the-Solent line. The view is looking towards Gosport with the old Garrison church on the extreme right. The 'up' side building is shown before it was extended whilst the signal was probably controlled from a lever at its base. *National Railway Museum*

the existing Anglesea level crossing. Approval for this was given, although for reasons that are unclear it would seem the work was never carried out. Surprisingly perhaps the request was also made to the South Western and no record of it appears in the Stokes Bay Company minutes.

In the meantime work on the new curved connection for Stokes Bay trains was progressing well, this having an 18-chain radius compared with the original 15-chain connection. The new line was brought into use on 1st June 1865 and was double track in identical fashion to the remainder of this route. On the same day the new station at 'Stoke Road' was opened, this coinciding with the introduction of the summer timetable whereby two trains each way took the new route. There was an immediate saving of 15 minutes in the journey time from Waterloo to Stokes Bay, although as time passed it was the norm to use the new spur more often in summer than in winter.

In later years it became practice to attach the coach from Stokes Bay to a train from Portsmouth at Fareham. Whilst in late 1875 the attachment was made at Bishopstoke instead. The best overall time from Ryde to Waterloo was three minutes under three hours. But such facilities were at the time only available to first and second class passengers, third class travellers having a journey of over four hours, although this was altered from September 1882 when South Western policy allowed such lowly individuals onto their fast trains.

As mentioned in a preceeding chapter, the electric telegraph had been installed on the main line at a very early stage. This was now being used to assist in the regulation of train services allied to the railway policemen already employed. The telegraph was extended to cover the Stokes Bay line as well sometime after 18th February 1864, although apart from knowing that signals were certainly in existance by 1865, detailed signalling arrangements of the period are somewhat obscure. It is likely also that any changes or improvements provided by the South Western consequent with the working of Stokes Bay trains were chargeable to the independent Stokes Bay Company, the telegraph then a typical example, although the cost of this is not recorded.

It is now time to look at the pooling arrangements applicable to the Portsmouth area and affecting the South Western and Brighton companies. In this arrangement both parties agreed to equal rates for traffic destined from their respective London termini to the port and therefore the competition which had at one time almost resulted in open warfare was avoided. The South Western's share of receipts was set at 60% of revenue, this recognising the latter companies routes to Portsmouth as likely to be those attracting most traffic. But to administer this agreement, which was in effect little more than a truce, the South Western had to agree to include the Gosport area in its contributions. Gosport itself, although providing a reasonable revenue return, already took second place to Portsmouth. After all was this not the original aim of the branch? Consequently the Gosport traffic and development of the Stokes Bay line was affected from the outset. The South Western was unable to route all Isle of Wight bookings via the new line even after it has been absorbed. Consequently, with the pooling arrangement, it mattered not how full the Stokes Bay trains were to the shareholders. This was a matter of often voiced criticism. As mentioned before, however, the Stokes Bay line was a valuable insurance to the heeding of the pooling agreement for both parties, something no doubt the Brighton were well aware of. Even so through bookings originating from places other than London and destined for the Isle of Wight, were often routed via Stokes Bay. One often quoted example is Banbury to Ryde, via Reading, Basingstoke and Stokes Bay.

The Royal Travellers

Despite the opening of the new line, the Queen and other Royal travellers remained loyal to the Clarence Yard station, three occasions in 1863 alone, when preference was given to the established route. Similarly, in later years, there is no record of the Stokes Bay line ever being used by the monarch. Even so local legends die hard, for certainly the royal baggage was routed over the new line joining the special trains at Basingstoke. Perhaps a fitting conclusion may be that even upon Victoria's death at Osborne in 1901, the route for the funeral train was still from Clarence Yard.

Before this, however, the Clarence Yard line was used many times. Each occasion demanded that elaborate arrangements be made. An example of this is the journey of Friday/Saturday 23-24th August 1878 when the Royal Train was to travel from Gosport to Dunbar, one of eighteen occasions when such a journey was made between 1873 and 1892.

For these trains, vehicles of London and North Western Railway origin were utilised and dictated the presence of that Company's Superintendant of the Line often accompanied by the Locomotive Superintendant as well. Once the Royal Yacht had docked, the first to alight were the household servants and ladies-in-waiting who made their way to vehicles at the front of the train. Each coach had written on the side the names of its occupants, with contemporary reports stating that;

> '......owing to the shortness of the platform, the vehicles in front in which they travel, have each to be served by a special set of portable steps to enable the passengers to enter the doors without 'clambering up'. These steps, however, were not there on this first occasion (1872) but the ready attention of Mr. Scott, one of the chief authorities of the dockyard, always provided them in subsequent years.'

This today seems to be a rather strange quote implying the Royal Train was unable to fit alongside what was a 520 foot platform.

It was the custom for the Queen, when disembarking, to be

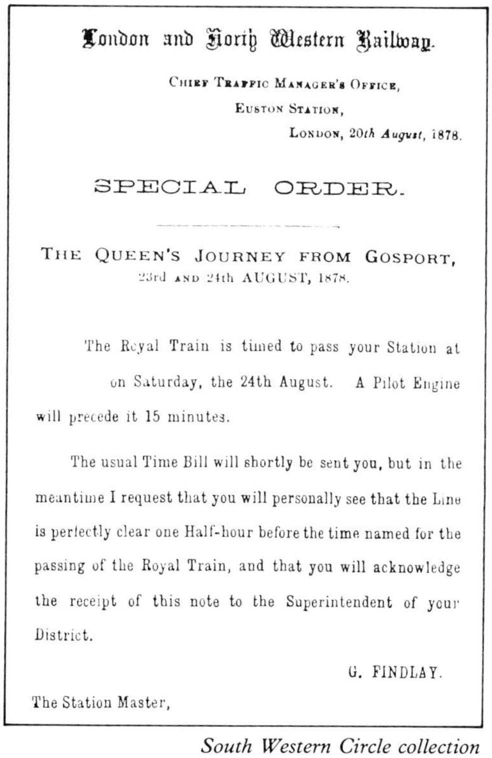
London and North Western Railway.

CHIEF TRAFFIC MANAGER'S OFFICE,
EUSTON STATION,
LONDON, 20th August, 1878.

SPECIAL ORDER.

THE QUEEN'S JOURNEY FROM GOSPORT,
23rd AND 24th AUGUST, 1878.

The Royal Train is timed to pass your Station at on Saturday, the 24th August. A Pilot Engine will precede it 15 minutes.

The usual Time Bill will shortly be sent you, but in the meantime I request that you will personally see that the Line is perfectly clear one Half-hour before the time named for the passing of the Royal Train, and that you will acknowledge the receipt of this note to the Superintendent of your District.

G. FINDLAY.

The Station Master,

South Western Circle collection

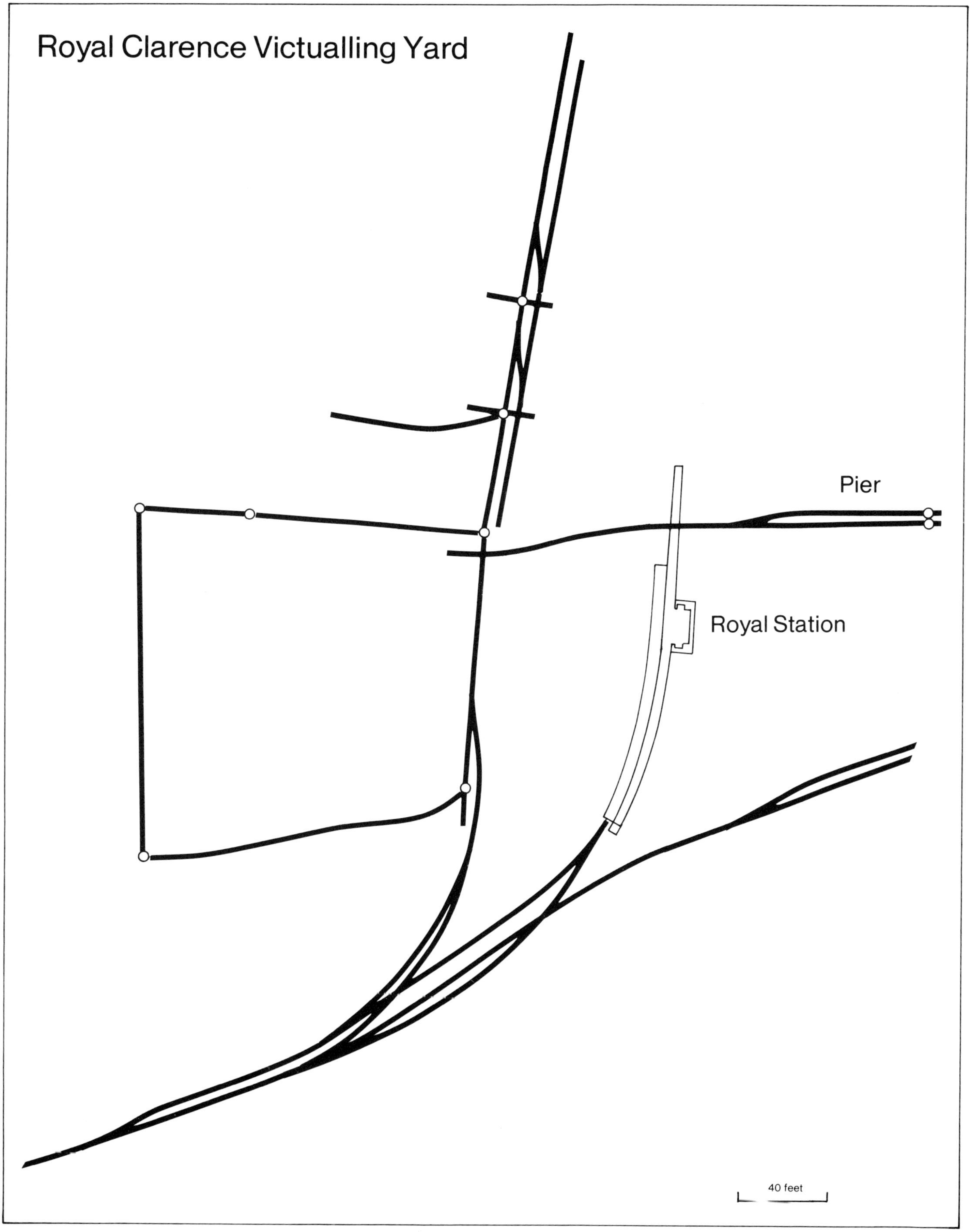
Royal Clarence Victualling Yard
Pier
Royal Station
40 feet

presented with the 'keys of Portsmouth', although she allowed these to be retained by the naval and military chiefs. The various equerries then escorted Victoria the short distance to the platform, at the same time emparting any special late requests to the railway staff. The royal saloons themselves were situated four or five vehicles from the end. It is reputed that no use was ever made of the private waiting rooms on the station itself.

The signal to start would be given to the rear guard by the Queen's personal servant, for some years the redoubtable John Brown. The rear guard then forwarded this to the front guard who in turn passed it to the driver. Both the former individuals resplendent in a uniform of scarlet and gold.

Upon leaving Clarence Yard the train would sometimes slow or stop outside the station where local children would line up to cheer and occasionally perform a well rehearsed recital. Such an event was normally undertaken on the 'down' journeys when it was quite common for Victoria herself to acknowledge their efforts from the saloon doors.

The Royal Trains were preceeded by a pilot engine, whilst in addition, the monarch insisted upon certain stringent 'safety precautions'. One of these was that branch lines were to be sealed off along the route. The Company complied by spiking the points and wedging the respective signals at danger. There was also a requirement that literally hundreds of gangers and pointsmen were posted to operate the hand signalling system. It would appear that Queen Victoria had a distinct mistrust for both branch lines and signalboxes.

The 'Hampshire Telegraph' provided a valuable insight into the working arrangements for the period:

> 'The Company's first duty was to print fifty copies of a complete 'way-book' of the journey plus a special copy for the Queen...... printed in purple on white satin, embossed with the Royal Arms and fringed with gold thread. It was laid down that the way-book should include a map of the full length of the line with the gradients shaded, the branch lines to be marked in red. Also incorporated in the way-book was the route schedule which had to show the exact time at which the Royal Train was due to pass each station or halt on the journey. By referring to the pendant watch on her bosum the Queen was able to check against any attempt to speed, ie. progress at more than 30 mph. (Other reports have widely stated Royal Trains of the period were permitted to run at 40 mph.) The Queen also required a scale plan of the entire train showing the exact number of the carriages and the dimensions of each and the names and precise seating situations of every occupant.'

Whether such precautions were ever put to the test is unknown, although one unfortunate event is recorded on 18th August 1875 but this time when the Royal Yacht *Alberta* was in collision with a schooner whilst en-route from Osborne. Unfortunately two persons from the schooner were thrown into the sea in the process, one later drowning. The Queen had apparently been witness to the whole trauma and upon arrival at Clarence Pier was reported as being 'very troubled'. It was even spoken of delaying the Royal Train, although after discussions this did in fact leave for its destination of Carlisle some 30 minutes late. This is one example where the generous schedule was used to advantage, for an 'on-time' arrival was later achieved.

Locomotives

Details of the first locomotives known to have used the line are sketchy in the extreme, although it is reasonable to conclude that various of the early South Western classes were employed on the services. It is recorded that a locomotive was stabled at Gosport from the earliest times, for many years this a member of the 'Nile' class of tank engine. Another locomotive known to have worked to Gosport was No. 114 *Frome* a 2-2-2 tender locomotive of Christie, Adams and Hill manufacture which hauled a royal train between London and Gosport in December 1861.

Records become clearer from the time of the use of the Beattie 'well' tanks. With numbers 248, 250, 253 and 255 working the services over both the main line and Stokes Bay branches up to the time of the demise of the class in the years prior to the turn of the century. These engines were ideally suited for the light

What may well be the earliest actual photograph of Gosport station taken around 1859. The engine is probably typical of those used on the line at the time. *Dennis Tillman collection*

trains then in operation, although with the proximity of the large depots of Northam and Fratton it is likely that additional motive power requirements were provided from these locations as required.

Accidents

Of all the surviving records relating to the railways in the Gosport area those covering the various accidents and incidents are perhaps the most interesting. The recorded ones are certainly no more or less eventful than those applying to almost all railways of the period. Of these perhaps the most intriguing is the first known of. Occurring on 23rd April 1856, 2-4-0 No. 123 *The Duke*, when on a Royal Train, over-ran the buffers at Gosport finishing up in the road. No serious damage was recorded, although a replacement locomotive had to be urgently sought for the return journey later that day. But this in itself poses several interesting questions. Was there a member of the royal family on board? Why refer to Gosport Station rather than Clarence Yard? Certainly it is unlikely the latter locality is actually the case, for at the end of the platform there is no road, just a short distance to the sea. Likewise, what were the turning arrangements for engines on RoyalTrains? Assuming of course No. 123 had worked down chimney first it would have run to Eastleigh for turning, as this was several years in advance of the triangle later provided.

There are two references to level crossing accidents, the first on 25th March 1860 when the minute book documents an;

'.....accident to a child on a level crossing near Gosport......'

and later on 4th July 1863 as follows:

> '........The railway gates at the road crossing of the Stokes Bay line, near Little Anglesea, were smashed through on Monday afternoon last, by one of the trains passing through them. Fortunately the accident was confined to the destruction of the gates........'

This was the first adverse incident recorded against the then almost new Stokes Bay line.

Later, on 28th September 1869, another incident is related;

> '.......line out of order at Brockhurst. 10/- gratuity to driver and 5/- to fireman for noticing the fault. 10/- fine to the ganger of the district.......'

Potentially the most devastating to date took place on 15th January 1884, when the 10.50 am 'up' train from Stokes Bay left the rails just north of Brockhurst station. The train was made up of tank engine No. 248 and two coaches; a six-wheeler upon which the body was hung by chains and a 4-wheeled brake van some 23′ 8″ long on a 10′ 6″ underframe. The train contained few passengers none of whom were seriously injured. Being a passenger train accident the Board of Trade conducted an enquiry, Col. Rich had some very pointed remarks to make over the running of such vehicles at the 'high speed of 45 mph.....'

It is known that at least one other accident happened, although details are again rather vague. This was on 23rd December 1884 when gatekeeper Vaughan was killed by a train believed to have been on the Stokes Bay line. Mrs. Vaughan later approached the South Western for financial assistance and they provided a gratuity of £10 in June of the following year.

As recorded above, the various level crossings had been the site of more than one incident and yet another became the worst, on 30th November 1895, at Cambridge Road when two men were killed whilst crossing the line. It transpired that there was no gateman in attendance between 7 pm and 8 am, the accident occurring between these times. But with a train service still running it is not surprising the Company came in for an amount of criticism. It is reported that by March 1896 a man was present for the whole service.

The final recorded event again concerns the Stokes Bay line and took place on 12th April 1899. Driver G. Cox was approaching the terminus on the 4.18 pm train from Eastleigh when he saw an item of luggage roll off the platform into his path. The train was successfully stopped, for which both Cox and his fireman were awarded an extra days pay.

Further Proposals

Despite the existence of what were now three separate methods of travel between the Isle of Wight and the mainland, (via Portsmouth, Southampton or Stokes Bay – the Lymington service was still in the future) the victorian entrepreneurs felt there were still further possibilities for expansion. This resulted in 1874 with the proposed Cowes and London District Railway. The intention was for a new line some three miles long from a junction with the Gosport line just south of the present Redlands Lane bridge to a terminus at a new pier at Hill Head. From that point it was stated to be but a short distance to Cowes. Railway passengers there then faced with a choice of routes to the capital travelling via either Bishopstoke or Havant. The promoters were successful in obtaining the necessary Act of Parliament, allowing three years for construction. But work was never started and today the whole scheme is recognised for what it was, a hopeless failure from the outset.

The decade of the 1870s also saw other changes taking place, later to have a profound affect upon the railway fortunes of the Gosport area. In 1876 the line had been extended at Portsmouth to the Harbour station from whence it was an easy step to board the vessel destined for the Island. Accordingly traffic on the Stokes Bay line commenced a slow decline, the services tending to be regarded as providing for local needs rather than catering for through travellers. Even so it was affectionately referred to as the 'Family Route' compared with the legend 'The Shortest and Most Direct Route' when speaking of the journey via Portsmouth. One may also ponder for a moment upon the fortunes of the line from the early days of the twentieth century, especially with the Meon Valley line now open and laid out for fast running. Why was there not a more determined effort to attract further trade? The internal politics of the railway companies is perhaps the only explanation.

New Works

Apart from those already mentioned, the years prior to the turn of the century were those when expansion rather than contraction were the order of the day. Some, in particular the new stations, have already been mentioned briefly, (these are dealt with in more detail in chapter seven), whilst others not of particularly public significance, may otherwise escape attention. To commence it is first appropriate to return again to the Fareham tunnels, as these required spasmodic attention almost from the time of opening and still a matter of concern to several locals. Such concern was hightened when passengers were able to view workmen with lengths of timber seemingly shoring up part of the roof space. Upon enquiry the official reason for this was '....whitewashing to be undertaken.....' further enquiries by the Board of Trade as a result of a number of complaints revealed that an extra course of bricks was in fact being added.

But this work had an obvious unfortunate side affect, that of reducing the clearances available. The result was a special notice issued with the Working Timetables of February 1874 as follows:

> 'Previous to a 'down' or 'up' goods train entering the Fareham tunnel, the off-side lamp must be removed. As soon as the 'down' or 'up' goods train has passed through the Fareham tunnel, the off-side lamp must be restored to its proper position. In the event of it being impossible for the guard or breaksman (sic) to get to the off-side lamp to remove it, while the train is in motion, its removal must take place at the previous station; that is the station at which the goods train last stops before entering the tunnel in either direction.'

Meanwhile, almost from the time of opening of the extension to Clarence Yard, this short line had been recognised as a valuable asset in the transportation of naval stores. Accordingly by the 1860s there existed a considerable network of sidings within the confines of the victualling yard, together with numerous wagon turntables. Perhaps as a result of this additional traffic now passing over the line, modifications were required and in June 1881 a Board of Trade inspection took place of some additional work on 'Her Majesty's Railway'. An iron bridge carrying the line across a ditch had been replaced by an embankment and wrought iron bridge weighing just under 8¾ tons, the new bridge providing for a 40 foot opening.

The L & SWR minutes for 26th April 1873 also provide a puzzling feature regarding the provision of a '....siding for coal traffic on the Stokes Bay line...', the obvious question is where, if anywhere, did such a facility exist? It is most likely that it was never provided. A few years later, on 10th October 1880, there is a reference to a, '.......cottage on the Stokes Bay line to be pulled down due to its delapidated condition....' but again no other details are recorded.

Tramway Competition

In December 1878 the first stirrings are recorded resulting in the Gosport Street Tramways Company promoting a system of tramways within the borough. The gauge was settled at 3 feet, whilst at a meeting of 1881 it was clearly stated to the shareholders that '......there were no intentions to extend the line to Fareham.....', this is despite the fact that powers for this had been obtained.

Construction began shortly afterwards with the new system commencing operations on 17th July 1882 between the town and Ann's Hill. That the service was an immediate success is gauged from the plans for northward expansion proposed only two months later. This was to take the trams to Brockhurst, the extended service beginning in February the following year. Of course at the time it was only a horse drawn service privately operated, take-over by the local corporation, electrification and subsequent extension to Fareham were to come after the turn of the century. The railway had the opportunity to compete on what were favourable terms.

So did the railway recognise the threat proposed? Probably not but if they did they failed to take it seriously for in 1889 there is a report of correspondence over the existing level of service between Gosport and London whilst a request for a station at Forton was declined. The train service at the time consisted of some seven passenger trains in each direction on weekdays. In addition there were four trains in either direction on Sundays. There were also four trains from and to Stokes Bay taking the new connecting curve, all of these calling at Gosport Road station (renamed from 8th November 1866). With one exception all these trains also called at Brockhurst, whilst the service was itself a mixture of local trains terminating at Fareham with others running through to Waterloo, the journey to the capital occupying some 3½ hours. The timetables also provide some details of some interesting services and facilities, such as the 6.55 am starting from Winchester running direct to Stokes Bay with connections to Ventnor where arrival was scheduled for 11.04 am. Likewise Fareham Junction was shown in the timetables as, 'For Lee-on-the-Solent', whilst Stokes Bay could reputedly handle 'Goods, Grain and General Merchandise'.

The final stage in the expansion of the network commenced in May 1894 with the opening of the Lee-on-the-Solent line from its junction with the main line at Brockhurst. Consequently several changes were made to the junction station, whilst a few years later the main line saw much traffic associated with the movement of men and materials destined for the Boer War.

Gosport High Street, complete with tram track. *Gosport Museum*

Chapter Five

Lee-on-the-Solent

On the coastline some four miles west of Gosport lies the area of Lee-on-the-Solent, today a popular location for yachtsmen and to an extent, bathers, whilst private accommodation commands a high value.

Its ancient history is in some ways similar to that of its neighbour, with flint implements of both paleolithic and neolithic periods uncovered at various sites in the area. Romano-British items of pottery are also to be found. By the middle ages the name of 'Leya' was being used, the later term 'Lee Britain' probably taken from the one-time owner of the estate Hamo Bruto, a character of Normal origins.

Apart from the digging of gravel there seems to have been little commercial development of the locality until the late 19th century, at which time the land around Lee was in the ownership of Sir John C. Robinson F.S.A., HM Surveyor of Pictures. Sir John did not himself live in the area instead having an estate at Newton Manor near Swanage, but saw in Lee the possibility to develop the locality into a 'watering place' similar to Bournemouth. Thus the location was showing the first signs of change towards the holiday resort for which it is renowned today.

The redevelopment of the site from its then description of a hamlet of but five cottages, commenced in 1885 under the direction of one of Sir John's sons. One of the first steps was to provide the Marine Parade, a 50 feet wide roadway more than a mile in length, parallel with the sea and separated from this by a grassy strip some 150 feet wide to be used as a park.

Shortly after this plans were laid down for the foundations of a new housing development, the whole neatly laid out with wide roads to provide access to the numerous new building plots. As far as the developers were concerned there remained only two further requirements to complete the project, a pier and a railway. The victorian generation's love of the former were manifest at numerous seaside watering places the length of the coast-line. Accordingly powers were obtained in 1885 for a 750 foot long monolith, to the design of Messrs. Galbraith and Church. The requisite £10,000 funding was supplied by Sir John Robinson. Work commenced almost at once and was completed less than three years later when on April 3rd 1888 Lady Robinson performed the opening ceremony. Within a short time a regular steamer service was plying between Lee and the Clarence Pier at Southsea. The new pier provided with seats, glazed shelters and of course a pavillion. Nearby a picturesque old farmhouse had been enlarged and converted into the 'well-appointed' Victoria Hotel.

The pier, Lee-on-the-Solent. *Gosport Museum*

But despite these improvements the ferry operators found it only viable to operate services in the summer months. Rail connection was seen as being essential to further promote the commercial development of Lee. It was thus a short step to the formation of the Lee-on-the-Solent Railway Company. The five original promoters were Charles Langley Whetham, James Edward Hunter, William Stephens Cross, Harry Emans Pollard and Charles Edward Morgan.

A survey by the engineer R.H. Tigg estimated the expense of the enterprise as follows:-

35,000 cu.yds of cutting through soft soil at 1/6d per yard.	2625
12,000 cu.yds of embankments and ballast	250
Accommodation bridges and works	1250
Sea wall	2000
Culverts and drains	250
Metalling of roads and level crossings	200
Gatekeepers houses at level crossings	300
Permanent way including fencing. 3m. 9.5ch. at £1760 per mile	5489
Permanent way for sidings	1320
Stations including engine and carriage sheds and signalling and telegraph work	5000
	£18684
10% contingencies	£1868 8s
	£22052 8s

Accordingly the first steps were taken to secure the necessary parliamentary powers, the company being successful in obtaining the required sanction on 14th April 1890. Little is known of the moves leading up to this consent but the first recorded meeting of the Lee-on-Solent Railway Company (the railway it appears not using the full title of the town), was held on 13th October 1890 at 263 The Strand, London. Nine persons were present, with Mr. Whetham in the chair, the post of chairman changing several times at ensuing meetings.

Contemporary reports also record a considerable amount of correspondence with the Board of Trade, local councils and Post Office. This was over the matter of the positioning of telegraph poles and possible severence of footpaths. In a lighter vein the anticipated charges for certain categories of traffic were published, passengers and horses at 2d per mile with compost and manure at 1¼d per mile!

It was proposed that a draft agreement be made with Messrs. David Laing & Sons for the construction and equipping of the railway. Then at the Board meeting of 11th November 1890 it is reported a Mr. F. Gilham was appointed engineer to the line at a salary of £500 p.a., the fate of Mr. Tigg being unclear. Similarly what happened next is unclear but certainly an agreement with Laing's was reached for the construction and worth £24,375. The contractor's 'subbing' to a Mr. Kingswell for work along the foreshore at Lee itself. Laing's reported that on 5th May 1891 they were in possession of all the requisite lands and that construction was estimated to take around seven months. However, the first of the difficulties began in April 1891 when Gilham wrote to the Board informing them he was leaving the country for 'some months'. Then on 17th November 1891 Gilham again wrote, this time tendering his resignation which was not accepted by the directors, although moving ahead in time slightly the minutes of the company record the appointment of a Mr. P.W. Meik as engineer on 27th February 1893.

Regretfully little is known of the contract between the railway company and Laings. Work certainly had not started by 17th November 1891 for on that day the secretary, Mr. Petter, was instructed to write to Laings '.....demanding work to be proceeded with at once.' Unusually, no parliamentary act authorising construction seems to have been required, the line instead being permitted under an order of the Railway Construction Facilities Act of 1864. One of the very few British railways built under such powers.

By early 1892 little work had been carried out. The directors visited their unfinished railway in May of that year followed by a meeting two months later at which the contractors were deemed as '.....equal to the task.....' of completing the works. A report was to be obtained (from the engineer? – if there was one) as to 'the cost of completing the railway'.

For the next few months matters are unclear, the only interesting feature was when it was reported that, '....proceedings are to be taken against three previous directors....' the reason for this is not stated but more than likely was due to financial mismanagement. One of those the subject of litigation was a former chairman of the railway! This left the railway company in a 'catch 22' situation, unable to arrange for completion of the line and with no interest shown by speculators in acquiring further shares.

What would seem to have been the next step, an approach to the South Western, was not in fact made and instead, in November 1892, Mr. Robinson junior, informed the board that his father was interested in seeing the project completed and might be able to render the necessary financial assistance. Such an offer was naturally ideal to the railway company, whilst from Sir John's viewpoint the railway was an essential part for the development of his Lee-on-the-Solent Estates Company. Matters between the parties were quickly settled. Sir John taking up what is believed to have been the remaining unsold shares of the £30,000 capital of the railway company.

It was now a question of obtaining a new contractor. An approach was made to Messrs. Pauling & Elliott of Westminster early in 1893 who agreed to take over and complete the unfinished works. This was achieved in the main without any major construction difficulties, although the 'boggy' nature of the soil near the bridge over the River Alver slowed the work somewhat. To assist in the work Pauling's brought with them at least two 'standard' locomotives, this being a usual feature of railway construction of the period; 0-6-0 saddle tank of Manning Wardle build, No. 156; 0-4-0 saddle tank of Manning Wardle build, No. 334 *Stanley*, whilst later an 0-4-0 contractors' engine is mentioned.

From various accounts of the time it would appear that the physical works were completed by the spring of 1893. The following description of the works was sent to the Board of Trade on 24th March 1893 as a prelude to the official inspection;

> 'Line to have 60lb rail fastened by spikes to cross sleepers. As the line is to be very light for some years after opening it is proposed to dispense with stations except at Lee terminus where the station is half complete and at Brockhurst where an adjoining platform will allow passengers to use the LSWR waiting room. It is proposed to work the railway with cars of the American type, having end platforms and low steps. The platforms to be only of a height sufficient for stock with steps. The level crossing gates to be hand-operated by a lever by the train conductor. The gate machinery would be constructed by a first class firm of signal contractors and will be well balanced. After the train is clear of the road the flanges of the wheels will actuate a draw-bar releasing the balance weights and the gates would again shut across the railway.'

This statement raises several interesting issues. It is the first time for example any reference has been made to a possible connection with the South Western at Brockhurst, some sources having suggested the line was to link in at Fareham, although this is believed to be an early error perpetuated over the years. Likewise up to now there has been no reference to any locomotives and rolling stock for the line. Perhaps the most interesting feature though is the suggestion within the document, in plan form, that four stopping places were to be provided, Elmore, where there was to be a passing loop, Browndown, Privett and Pound Lane Crossing. Each of the three last named would have had a level crossing of the type mentioned, similarly each having a platform either side of the crossing. In this way trains would stop at the platform on the approach side of the crossing whereupon the guard would open the gates across the road leaving the gates to be restored across the railway by the actuating mechanism. (Plans of the proposed station layouts are shown in chapter seven.)

Such a proposal was of course a radical departure from the normal practice and there followed a not inconsiderable degree of correspondence between the Lee-on-Solent company and Board of Trade. It transpired that the Board of Trade would not agree to an official inspection of the works until the matter in question was resolved.

Some agreement, however, must have been reached for on 15th July 1893 the line was visited by Major Yorke, who had already been briefed by Major General Hutchinson that gate-keepers boxes had not been erected as it was the Company's intention to work the complete line as a tramroad. (The original engineer's estimate of 1889 does of course mention gatekeepers houses for the level crossings, one can only assume that the Lee-on-Solent company wished to dispense with the need for these on the grounds of reducing costs.)

Major Yorke's inspection reads as follows;

> 'I have the honour to report .... that I have this day inspected the Lee-on-Solent Railway ..... The line is single of 3m. 9½ch. long and the company have purchased sufficient land for doubling if required. The steepest gradient is 1 in 66 and the sharpest curve of 10ch. radius. The cuttings and embankments are unimportant. The line is fenced with post and rail. Flat bottom steel rail of 60lb per linear yard has been used, secured by dog spikes. The sleepers are of creosoted baltic timber 8′ 11″ x 9′ x 4½″. There are 11 sleepers to a 30′ length on the straight track and 12 on curves. All points and joints bearing plates below each rail are secured with 4 spikes. The gravel ballast is sufficient except in the vicinity of the River Alver where more is required. The line has 3 underbridges, 2 of 12′ and one over the River Alver of 20′. All are of rolled steel and resting on timber piles. The latter sunk to a depth of 26′ for the smaller and 46′ for the larger bridge. Deflection tests were made with a 4-wheeled contractor's locomotive believed to weigh approximately 15 tons. There are 2 level crossings over public and one over war department roads. All have gates and wickets but the gates require lamps for night use. The rail width available between the level crossing gate is 11′ 3″ which is sufficient if tramway stock is to be used. There are 4″ high platforms at Elmore and Lee without shelters or booking offices. Buildings at Lee noted as under construction. No signals at all. All sidings being locked with a key on the train-staff, at the time of inspection, however, the sidings are locked by a loose key. No facing point locks provided and no check rails on 10ch curves.'

The report concluded that the line failed inspection on no less than thirteen points, including that gauge ties were required on any curves of less than 15ch radius. The comment was also made that the line appeared to have been partly built outside the limits of deviation set and that the gradients were steeper than those originally proposed. Major Yorke sent a telegram to the Board of Trade stating, '..... I cannot recommend the opening as the works are incomplete'.

What then was the real situation? For if the Inspecting Officer's report is to be believed and they are usually very accurate, there are now several contencious issues compared with those previously reported and in addition many raised questions. Why for example was the line built outside its legal limits and what of the mention of the 10ch. curve at Elmore when the original plans show this as 5ch? There is also no mention of the twin platforms and spring-operated level crossing gates or for that matter the existence of four intermediate stopping places. One would expect to find the answers to these issues in the report of the next directors meeting, but tantilisingly that is not the case. The subject did not even attract comment of any sort. So why then did the inspection ever take place? Possibly the answer is simple to find as a desire by the directors to open their railway as soon as possible in whatever form they could and so attract some return on the investment was of major importance to them. Pressure was no doubt exercised by Sir John Robinson whose prime object had been for improved communications for Lee-on-Solent itself.

There now follows a period of unrecorded activity, during which time arrangements were made to bring the line up to a suitable state to be sanctioned for carrying passenger trains. Meik did, however, attempt once more to allow the line to be opened, in its existing state, by suggesting to the Board of Trade it be classed as a 'tramroad' – a blunt refusal was the answer obtained.

By early 1894 correspondence is recorded between the South Western, Lee-on-Solent company and Board of Trade over the subject of through trains, the Board of Trade requiring different arrangements at Brockhurst should such services be contemplated. In addition standard height platforms would be needed at the stopping places, whilst a porter and waiting shelter would have to be provided. Records regarding the South Western show them to be a passive observer and refusing to be drawn into the discussions.

But at last the changes had been completed and on 28th March 1894 a letter was sent by the Lee-on-Solent board to the Railway Inspectorate informing them the line would be ready for re-inspection the next month. This was followed on 7th April by a further letter stating the intention to open the line on 28th April 1894. Several other interesting points being mentioned;

> '.....we will use the two engines belonging to our company and the two American or tramcar-type carriages have been fitted with continuous footboards at the same height as ordinary carriages. Platform heights have been raised to 3′ and will be provided with lamps. If permission is given to using ordinary passenger stock in an emergency, an undertaking will be given not to stop at other than at Brockhurst and Lee.'

One may again ponder as to the description and details of the locomotives and rolling stock referred to but these are discussed in detail later.

An early postcard view of Lee showing the station buildings to advantage, the engine apparently in the process of running round its train.
Lens of Sutton

But perhaps significantly it may be appropriate to quote from the final paragraph of the report, which added;

'.....if the company had known all these obstacles were to be placed in their way ...... would either have abandoned the project or converted to a tramroad very early on.'

The re-inspection, again by Major Yorke, took place on 7th May 1894, his report submitted in the usual form, with the following summary of his comments;

'..... since the time of the original inspection the 20' bridge over the River Alver has been abolished – the river deviated ...... because of settling occurring on this new embankment which is on marshy ground it should be carefully watched ...... There are now small corrugated iron gatekeeper's cabins at each level crossing, each with a clock, stove, seat and table. A man to be provided at each site ..... now additional stations at Privett and Browndown ..... it is no longer the intention to use Elmore as a stopping place ..... Lee Station is ..... comfortable and convenient.....
However still required:

1. Buffers at Brockhurst on the siding between the LSWR and Lee-on-Solent Co.
2. Platform ramps are too steep.
3. Fencing needed at the rear of Privett and Browndown stations.
4. A 10 mph maximum speed as agreed by the Company.

Provisional sanction given for the opening with a further re-inspection in June.'

Accordingly at last the line could carry passengers, the Company wasting no time and commencing a service, allied to the usual formal opening on 12th May 1894. The ceremony was performed by Lady Clanwilliam, accompanied by the usual group of dignatories. (A contemporary report in *The Railway World* for July 1894 suggests the formal opening was not until 31st May but there is no other evidence to support this.)

Returning again briefly to the circumstances of 1893/4. It is by no means clear that the contractors Pauling and Elliot were responsible for the changes necessary to allow for opening of the line following the first inspection by Major Yorke, for records show the firm of John Price involved in a railway construction contract in the Gosport area c. 1894. Little is recorded against this work, apart from the fact that three locomotives of standard gauge were utilised, one of which was a small Manning Wardle saddle tank named *Gosport*.

In anticipation of regulating services, the directors had, in August 1893, formed a Traffic Committee consisting of two of their number. In addition there was a manager, Mr. E.B. Ivatts, later replaced by one of Sir John Robinson's sons, Edmund. It is not clear if the post also involved a seat on the Traffic Committee mentioned.

To operate the new service a locomotive was hired from the South Western. The previous reference made by the Lee-on-Solent Company to owning their own engines being somewhat ambiguous. Accordingly arrangements were made for this loan on an initial six-month basis, the first locomotive involved being a 2-4-0 side tank No. 21, and named *Scott*, this later alternating with another equally small machine. Two carriages were used, these of the 'tramway' type and comprising both first and second class accommodation. The Company had no goods stock,

LSWR No. 392, one of the Manning Wardle tanks alongside the branch platform at Fort Brockhurst during the period 1898-1903. *Lens of Sutton*

although mention is made of a passenger brake van, – this being a former LSWR vehicle but again details are unclear. (Full details of the locomotives and carriages are in Appendix I.

The initial service comprised eight trains each way daily with three on Sundays, all services calling by request at the two intermediate halts of Browndown and Privett. Later the timetable was changed to afford a slightly reduced number during the winter months. The fact that the first train did not leave Brockhurst until 10.15 am is a reflection upon the type of patronage the line wished to attract. With the directors and shareholders no doubt hopeful that they could at last expect a return upon capital the first figures were eagerly awaited. These showed that from the time of opening up to and including the 20th May 1894 total traffic receipts were a paultry £30 3s 5d, hardly enough to sustain expenses! Out of this had to come wages for various staff, the stationmaster at Lee alone receiving a salary of £5 19s 9d monthly. In addition there were ten other employees, commanding a wage bill of £39 10s 6d, one of whom was described as being a Policeman whose function it had been for several years to safeguard the line at the time when it had been under construction and also when dormant.

Certainly amongst the directors there was still great confidence placed in the enterprise, the hope being that through traffic could be enticed from the South Western. Over the years, however, there is little firm evidence to support the fact that passengers were able to travel direct to Lee-on-the-Solent without alighting briefly at Brockhurst. The track layout did not allow the running of through trains even if the vehicles were shunted empty across to the Lee line platform. Certainly excursion tickets were available, one recently recalled being a Sunday School outing from Bishop's Waltham direct to the Lee terminus.

The figures to the end of the first year of operation are of particular interest, showing a deficit to the South Western. This was to become a regular feature of the operation throughout the time the line was under independent control.

To Year Ending 31.12.1895.		£ s d
Loco Power: Hire of Engine		23 5 0
Oil and Stores		2 6
Passenger Receipts to LSWR		5 16 9
Parcels Receipts to LSWR		3 11 11
Bookings at Fort Brockhurst		16 8
Rent of Junction		20 0 0
Goods Account Paid to LSWR	Nov	17 1 4
	Dec	37 2 7
Printing Tickets		1 3 8
		£109 0s 5d

		£ s d
Passenger Receipts from LSWR		1 9 9
Parcels Receipts from LSWR		1 9 9
Goods Receipts from LSWR	Nov	7 18 9
	Dec	6 7 3
Balance Due to LSWR		84 8 11
		£109 0s 5d

Lee-on-the-Solent Railway

TRAIN SERVICE BETWEEN
Lee-on-the-Solent & Fort Brockhurst Junction (L. & S.W.R.)

1st JUNE to 30th SEPTEMBER, 1908 (or until further notice).

FROM LEE-ON-THE-SOLENT

STATIONS		WEEK DAYS.								SUNDAYS.				
		a.m.	a.m.	a.m.	p.m.	p.m.	p.m	p.m.	p.m.	p.m.	p.m	p.m.	p.m.	p.m.
Lee-on-the-Solent ...	dep.	9.20	10.35	11.50	2.10	2.50	4.10	6.30	7.30	3.0	4.35	...	6.40	
Browndown ..	..	...	Stop	by	sig	nal			...	Stop	by	sig	nal	
Privett	..	...		do.	do.				...	...	do	do.		
Fort Brockhurst Junction	arr.	9.35	10.50	12. 5	2.25	3. 5	4.25	6.45	7.45	3.15	4.50	...	6.55	
Fort Brockhurst Junction	dep.	9.45	10.54	12.11	2.39	3.29	4.46	7,19	8.49	3.24	5.4		7. 9	
Fareham	arr.	9.51	11. 0	12.17	2.45	3.35	4.52	7.25	8.57	3.30	5.10		7 15	
Eastleigh	,,	10.24	11.24	12.43	3.14	...	5.18	7.58	9.33	...	5.43		8 0	
Salisbury	,,	11.29	...	2.8	4.23	4.57	6.34	9.39	...	5.17	..		9.19	
Southampton	,,	10.46	11.49	1.c9	3c29	4c10	5 42	8.28	10.19	...	6.17		8.25	
Winchester	,,	10.54	11.53	1. 6	3.43	..	5.47	8.31	10. 9	...	6.19		8.34	
Basingstoke	,,	11.26	12.25	1.36	4.19	...	6.25	9. 1	10 39	...	6.56		9.6	
London (Waterloo)	,,	12.28	1. 31	2.45	5.40	7A 1	7.31	10. 5	11.48		8.40		9A59	
Fort Brockhurst Junction ..	dep.	10.12	11.27	12.24	3.10	4.19	4.44	6.50	8. 5	4.2	5.37	...	7.47	
Gosport	arr.	10.15		12.27	3.13	4.22	...	6.53	8. 8	4.5	5.40	...	7.50	
Gosport Road (Alverstoke)	,,	10.28	11.31	...	...	...	4.48	...	...	..	...	..	...	
Stokes Bay	,,	10.32	11.35	...	...	...	4.53	...	...	..	...	...	...	
Ryde Pier Head.. ...	,,	11. 5	12. 0	...	...	...	5.25	...	...	..	..	...	...	

TO LEE-ON-THE-SOLENT

STATIONS		WEEK DAYS									SUNDAYS.			
		a.m.		a.m.	a.m.	a.m.	p.m.	p.m.	p.m.	p.m.		a.m.		p.m.
London (Waterloo) ...	dep.	7a10	...	7.40	9.20	11.40	...	2.20	4a12	5a30	...	10.0	...	...
Basingstoke	,,	7.50	...	9.32	10.41	12.46	..	3.33	4.37	6.12	...	11.42	...	..
Winchester	,,	8.34	...	10.3	11.11	1.15		4. 2	5.8	6.45	...	12.15	..	5.30
Southampton	,,	8.50	..	10.28	11.50	12.55	2.15	3.50	5.50	6.20	...	3c0		5 38
Salisbury	,,	7.47	...	8 50	10.59	12.42	1.12	3.20	4.22	6.33	...	2.15		4.59
Eastleigh	,,	9.12	...	10.45	12.15	1.30	...	4.16	5.40	7.21	...	12.50	...	6.5
Fareham	,,	10.5	..	11.20	12.38	2.3	3.3	4.37	6.43	7.58		3.55	..	6.33
Fort Brockhurst Junction...	arr.	10.11	..	11.26	12.44	2.9	3.9	4.43	6.49	8.4	...	4.1	...	6.39
Ryde Pier Head	dep.	9.8		10.10	11.15		2.0	4.0	...	6.15				
Stokes Bay ..	,,	9.35	...	10.45	11.53		2.30	4.37	5.15	7. 0	...	..	..	...
Gosport Road (Alverstoke)	,,	9.39	..	10.49	11.57		2.34	4.41	5.19	7. 4			...	..
Gosport	,,	10.5	...	..	12.7	2.5	3.25	..	5.50	7.15	.	3.20	...	6.10
Fort Brockhurst Junction...	arr.	10.8	...	10.53	12.10	2.8	3.28	4,45	5.53	7.18	...	3.23	..	6.13

	a.m,	a.m.	a.m.	a.m.	p.m.	p.m.	p.m.	p.m.	p.m.	p.m.	p.m.	p.m.	p.m.	p m.
Fort Brockhurst Junction	dep.	10.15	...	11.30	12.55	2.30	3 50	4.55	7. 5	8. 5		4. 5	4.55	7. 0
Privett			..	...	Stop		by	sig	nal		Stop	by	sig	nal
Browndown ...			...	...		do.	do.	..	...	...	...	do.	do.	...
Lee-on-the-Solent .	arr.	10.30	.	11.45	1.10	2.45	4. 5	5.10	7.20	8 20		4.20	5.10	7.15

C Southampton West. A London Passengers by these Trains, travel via. Meon Valley Line.

NOTICE.—The published Time Tables of the Lee-on-the-Solent Railway Company are only intended to fix the time before which the trains will not start, and the Company do not undertake that the trains shall start or arrive at the time specified in the Tables, nor do they guarantee the connection of trains at the various Junctions. The Company give notice that they will not be responsible for any loss, inconvenience, or expense which may arise from delay or detention, or from non-correspondence of trains at the Junctions. The times of arrival and departure of the L. & S.W. Trains at Fort Brockhurst & other Stations are given for information only & the Lee-on-the-Solent Company do not hold themselves responsible for the accuracy of same.

FARES:—

From Fort Brockhurst to Lee-on-the-Solent & vice versa—1st Class 9d., 2nd Class 6d., 3rd Class 3d.

Through Day Return Tickets are issued from Lee-on-the-Solent to Portsmouth and vice versa, (via Floating Bridge Co. and Tramway to Station Road, Brockhurst; Passengers should allow 20 minutes from Gosport Hard)—Fares 1st Class 1/11; 2nd Class 1/5; Third Class **10d.**

Passengers can also book through to and from Principal L. & S. W. Stations.

E. A. ROBINSON, **Manager.**

H. W. Duffett, Printer, High Street, Fareham.

The terminal station at Lee soon after opening. *Railway World*

But despite the setback caused by the poor results, the directors were far from being disappointed. They attempted to 'sell' their line far and wide, this being detailed in a reported agreement with a Mr. Lawton for bill-posting timetables for the sum of £2 2s 10d.

The next set of figures show outgoings for the first six months of 1896 to be a little better.

For 6 Months Ending 30.6.1896.	£	s	d
Wages	237	3	1
Station Master's salary	35	19	0
Coal	48	12	6
Ticket printing		14	6
Uniform caps		13	10

One may perhaps wonder how such a situation could be tolerated and with the LSWR obviously aware of the matter, it comes as no surprise to learn that on 19th February 1896 the subject of a possible takeover by the large concern was discussed. Evidently no decision was reached, although the matter had gone as high as the South Western Board. Interestingly there is no mention of this move in the Lee-on-Solent Company minutes and no mention of a possible takeover until some twelve years later in 1908. With hindsight it is interesting to speculate on the situation. Even at this early stage the railway was of the type the redoubtable Col. Stephens would manage and certainly this person already had interests closeby with the Selsey Tramway.

Returning again to 1894, a re-inspection of the facilities took place on 5th November, this of course had been promised when provisional opening sanction had been given. As before, the Inspecting Officer was Col. Yorke (the change of title indicating promotion in the previous 6 months) he noted that;

> '.....full opening sanctioned .....all pending work has been carried out.'

But there was on proviso,

> '.....the banks of the two bridges should be watched.....'

Unfortunately failing to mention which two! One may, however, assume one of these to have been the River Alver bridge where the contractors record having to tip thousands of tons of soil until terra-firma was at last reached.

There was further correspondence with the Board of Trade the following year over the subject of continuous brakes on certain of the carriages. The Lee-on-Solent Company requested an increase in the speed allowed from the authorised 15 mph to 20 mph on the basis that all carriages now had this rudimentary necessity. One may perhaps ponder on the previous situation! The reply from the Inspectorate was favourable, the only condition being that gauge ties were to be fitted on certain of the curves. It should be mentioned that the line speed was later raised to 25 mph, the standard limit for 'light railways'. The Lee-on-the-Solent line later came into this category, although its method of utilising various complex powers under certain regulations and acts was at the very least, unusual.

By 1898 the finances had slightly improved, the maximum profit for any year recorded as £1,400, although the date of this is unknown. Perhaps then on the basis of their new found wealth the Company are recorded as contemplating purchase of their own locomotive, although it was later decided not to proceed with the matter.

Surprisingly though with all the trials of hardship involving the little company, relations with the South Western remained most cordial. At a meeting held on 17th December 1903 for instance it was reported the LSWR General Manager had travelled over the Lee-on-Solent branch and agreed to advertise the line in both the companies timetables and also on photographs in carriages.

The next stage in the story comes in March 1907 when an approach was made to the Board of Trade for permission to use 4-wheeled locomotives '....as traffic was very small....' The Inspectorate considered such a move 'inadvisable' but could not actually prevent it, a surprising comment from both sides especially when it is recalled a 2-4-0 tank had been in use from the time of opening.

There was a blow to the little concern in June the following year when the South Western informed them that both the locomotives they were hiring were worn out and no other suitable machines were available for the lines 8-ton axle loading. As before, however, the South Western were quick to add that help could still be provided, this time by operating the service with railmotors, the easy gradients and generous timings unlikely to show up the deficiencies in pulling power these vehicles had found elsewhere.

But before such a change could come about there had to be some legal changes, the LSWR officially taking powers to work the line as from 26th July 1909. The railmotor service started from 1st August, with the fate of the Lee line's rolling stock uncertain. What is certain is the vehicle concerned was No. 9 of the 'H13' class, sister machine No. 10 joining in from 1st September 1909.

LEE-ON-THE-SOLENT BRANCH.
RAIL MOTOR CAR SERVICE, First and Third Classes only.

	WEEK DAYS.												SUNDAYS.—(Commencing 5th July.)						
	a.m.	a.m.	a.m.	a.m.	a.m.	p.m.	p.m.	p.m.	p.m.	p.m.			a.m.	p.m.	p.m.	p.m.	p.m.		
LONDON (Waterloo) ... dep.	5 35	7J10	8 55	10 15	11 40	...	2 20	...	4J12	5J30	...	...	10 5	...	...	...	...	...	
FORT BROCKHURST dep.	8 55	10 15	11 30	12 40	2 40	3 30	5 4	6 10	6 55	8 15	...	...	2 42	3 30	4 12	5 40	6 20	...	No Service on Sundays during June.
Fort Gomer Halt... ...... „	8 59	10 19	11 34	12 44	2 44	3 34	5 8	6 14	6 59	8 19	...	...	2 46	3 34	4 16	5 44	6 24	...	
Browndown Halt............ „	9 2	10 22	11 37	12 47	2 47	3 37	5 11	6 17	7 2	8 22	...	...	2 49	3 37	4 19	5 47	6 2[illegible]	...	
Elmore Halt ... . „	9 5	10 25	11 40	12 50	2 50	3 40	5 14	6 20	7 5	8 25	...	...	2 52	3 40	4 22	5 50	6 [illegible]0	...	
LEE-ON-THE-SOLENT arr.	9 7	10 27	11 42	12 52	2 52	3 42	5 16	6 22	7 7	8 27	...	...	2 54	3 42	4 24	5 52	6 32	...	

	WEEK DAYS.												SUNDAYS.—(Commencing 5th July.)						
	a.m.	a.m.	a.m.	p.m.	p.m.	p.m.	p.m.	p.m.	p.m.	p.m.			p.m.	p.m.	p.m.	p.m.	p.m.		
LEE-ON-THE-SOLENT dep.	9 18	10 38	11 53	2 10	3 5	4 18	5 43	6 33	7 47	8 35	...	...	3 5	3 50	4 45	6 0	6 50	...	No Service on Sundays during June.
Elmore Halt.................. „	9 20	10 40	11 55	2 12	3 7	4 20	5 45	6 35	7 49	8 37	...	...	3 7	3 52	4 47	6 2	6 52	...	
Browndown Halt........... „	9 23	10 43	11 58	2 15	3 10	4 23	5 48	6 38	7 52	8 40	...	...	3 10	3 55	4 50	6 5	6 55	...	
Fort Gomer Halt........... „	9 26	10 46	12 1	2 18	3 13	4 26	5 51	6 41	7 55	8 43	...	...	3 13	3 58	4 53	6 [illegible]	6 58	...	
FORT BROCKHURST arr.	9 30	10 50	12 5	2 22	3 17	4 30	5 55	6 45	7 59	8 47	...	...	3 17	4 2	4 57	6 12	7 2	...	
LONDON (Waterloo) ... arr.	12 30	1 58	2 45	5 39	7J6	7 31	...	10 5	...	12 4	...	...	...	...	8 6	...	10J9	...	

J Via Meon Valley Line.

The spartan yet functional interior of the early rail motors, 2nd class passengers having to use wooden seats. *R. Curl collection*

The service was now ten trains each way daily on weekdays, with five on Sundays. These latter runs not starting until the 2.42pm departure from Brockhurst and running in July only. As before the arrangements for goods traffic is uncertain.

The South Western had thus taken a definite hold on Lee-on-the-Solent operations, although not to the extent of a full takeover. Instead they exerted their influence in other ways such as insisting on a name change for Privett Halt to that of Fort Gomer and so avoid confusion with the station on their Meon Valley route. A new halt was also opened at Elmore on 11th April 1910, how much this was due to the influence of the new operators is uncertain.

Possibly the introduction of the railmotors had an unexpected bonus, for in November 1910 the Gosport and Fareham Tramway Company applied for powers to extend the existing electric tramway network some 2½ miles from Bury Cross to Lee-on-the-Solent. Fortunately for the railway the extension was never built. Had it been, there would have resulted the unusual sight of rail and tramway running parallel for part of the journey. In the event competition came from a motor bus over basically the same route, this itself subsequently altered to compete directly from Brockhurst to the Marine Parade, Lee.

The railmotors lasted in use until 1915, when the working reverted to locomotive haulage and 'push-pull' trains, destined to remain the pattern for most of the remaining life of the line.

A description of the working in the immediate post World War 1 years is provided within a quote in G.A. Allcock's book;

> 'The train would travel alongside the Military Road as far as Gomer Halt, ...the area was often in flood and railway engineers had great difficulty in crossing the bed of the River Alver. Browndown Halt was situated at the beginning of Portsmouth Road. It was just a bare platform but actually had a stationmaster who was in charge of the crossing gates. The track proceeded along Portsmouth Road.... which was a gravel lane. Elmore Halt.... had a cutting with a board marked 'WHISTLE', a level crossing and a wicket gate, a small platform and a shelter with a nameboard. Across the road was a baker's shop where one could buy a delicious iced bun to munch whilst sitting on the fence waiting for the train. From Elmore Halt to Lee-on-Solent was a straight run.... on arrival the train crew would shunt back to the water tower to fill up the engine tanks and then disappear to the Working Men's Club, which was a former chapel, to refuel themselves before the journey back to Fort Brockhurst. The Lee-on-the-Solent Railway was advertised as taking passengers 'to where the rainbow ends.'

Thus the little company faced the advent of the Railways Act of 1921 threatening to force it to be amalgamated into the new Southern Railway. But before this could happen the finances of the Lee concern had to be resolved. An ever increasing deficit accumulated on certain of the share issues. At the time all £30,000 of the ordinary shares had been taken up with only £20 unpaid. Whilst in addition, out of an authorised borrowing capacity of £10,000, the requirement had only been to borrow £2,500. On the face of it not a bad achievement especially when considering the problems in construction and obtaining sanction to operate the line.

But against this must be stated the value of the line in so far as its assets were concerned. A mere £4,400 was set against creditors of some £14,600. Small wonder then the Southern were far from keen in acquiring a concern which would also make it then liable for its debts. The debate raged on long and fierce, the Lee directors seizing the opportunity to release themselves of what was an ever growing burden.

Eventually the matter was decided by arbitration at a hearing of the Railways Amalgamation Tribunal on 4th January 1923. The chairman agreed that the Southern should be responsible and an appeal by the Southern a few months later against the decision was dismissed.

The diminutive Lee-on-Solent Company had won, their last board meeting held on 1st February 1923 formally approved the takeover. The fortunes of the railway were now out of their hands and the impression was that they were well glad to be rid of it.

Chapter Six

Along the Line

Trains bound for Gosport from Fareham invariably began their journey from Platform 1, the track layout and position of Fareham station signifying the course of the original route as that to Gosport, the Portsmouth connection visibly a later addition. This legacy was perpetuated in modern times for despite the slewing of the Portsmouth lines for a more gradual curve to allow for higher speeds, it is still a cruel restriction to train timings. Had the original plans for a railway to Portsmouth come about in the 1830s then this would have taken the line on a more northerly course. Fareham instead being left totally on the line of rails leading to Gosport.

Leaving the station the line passed over the main A27 south coast road by twin brick arches. This was bridge number 28 from Eastleigh, the numbering continuing southwards and taking in the Stokes Bay line as well. Bridge 28 was for many years a bottleneck hampering the passage of high vehicles. The Portsmouth line then diverges to the left, check rails against the running line an indication as to the severity of the curve. Fareham West signalbox is then passed to the right, which was of basically the same design as the East box and controlling the lines at the south end of the station as well as the sections to Porchester and Fort Brockhurst. (South Western practice was to identify signalboxes by their proximity to London, hence East was nearer to London than West, even if geographically it would have made more sense to say north and south.) From 1934 onwards trains would collect a single line tablet from the signalman, this was the authority to proceed onto the section safe in the knowledge that no other train could be there at the same time.

The line is at this time running on embankments and parallel with a long siding on the site of the former 'down' line. This continues as far as Redlands Lane underbridge, a narrow brick arch having a 15 foot span. A little further on the right there is the site of the former Wych Lane sandpit. A narrow gauge line running within the excavations and, it is said, connecting with a private siding near to Wych Lane underbridge sometime during the latter part of the last century. Its exact course is, however, unknown, for it does not appear on OS sheets of the period and all traces have long since been obliterated.

The embankment has continued almost unabated up to this point and so allows a clear view to the right of the remains of Fort Fareham with open country beyond in the direction of Stubbington. To the left though the urban sprawl of Hoeford is visible, development soon encroaching on both sides, a mixture of naval, industrial and domestic premises. Since the problems reported with stability on this stretch in 1840 no further difficulties were reported, the rails on a falling gradient of 1 in 466

'T9' class No.30709 on a Gosport train at Fareham. *D. Callender*

A Hants and Dorset 'Bristol' passing under the twin arches at Fareham. Light shining on the front of the vehicle indicates the gap between the bridges. *Dennis Tillman collection*

since leaving Fareham with the embankment now slowly subsiding. The line becomes level at milepost 87 this continuing all the way to the Gosport terminus.

A little further on to the left is Bedenham siding. This is first mentioned in the minute book of 3rd March 1910 when the South Western agreed to rail connection linking the Bedenham Magazine depot with the main line. Cost of the connection was put at £638, this referring to the main line link only and not of course the railway within the naval depot. The Admiralty was to pay 10% of the expenditure for 25 years. Work started soon afterwards, for on 27th April 1911 the Board of Trade were approached for permission to operate the new siding prior to official inspection. Approval was required as it joined into a passenger line. This was granted with Major Pringle subsequently visiting the new work on 20th January 1912 when full sanction was given.

Meanwhile traffic had begun to Bedenham from about June 1911, access provided from the 'down' line only and although Fareham was in charge of the working, all outward traffic had to be worked to either Fort Brockhurst or Gosport before heading back towards Fareham. The siding was controlled during double line days by a single lever ground frame, release from this obtained from the signalman at Fareham West.

From the main line the siding, which at this point comprised of standard L & SWR fittings, ran round in a semi-circular south easterly direction, with a level crossing over the main Fareham – Gosport road, before entering the naval premises. The original layout within the yard is uncertain but what is known is the lines were extended about March 1914 to a Magazine Depot at Priddy's Hard, still within the complex. By 1918 the then layout was as shown on the plan, the various level crossings an operating difficulty to the transference of wagons from one site to another. After the singling of the main line a two-lever ground frame was substituted. This, as would be expected, was locked by the single line tablet, (later token) held in Fareham West box. Special transfer workings were provided between Fareham and Bedenham as required. There was also a narrow gauge system within Priddy's yard itself, the locomotives from this and the standard gauge network, together with some notes on the methods of working are detailed in Appendix II.

It is now only a matter of a half a mile or so to Fort Brockhurst station. This was approached in a cutting with Rowner Arch above carrying a road from the main A32 into what is now the Rowner estate. The cutting was the site in 1881 of a train being totally snowed in whilst on its journey from Fareham, a large contingent of marines from the barracks at Forton called upon to assist the railway staff in extricating both passengers and stock.

On the left-hand side was the 'down' distant signal next to which had at one time existed a wooden fogmans hut. This was the site of a fire on 25th April 1914, the cause of which was unknown, whilst the hut itself had to be demolished to extinguish the flames.

The crew of 'Q' class No.30536 giving up the single line tablet at Fareham West on a Gosport – Eastleigh goods working. *D. Callender*

BEDENHAM & PRIDDYS HARD RAILWAY

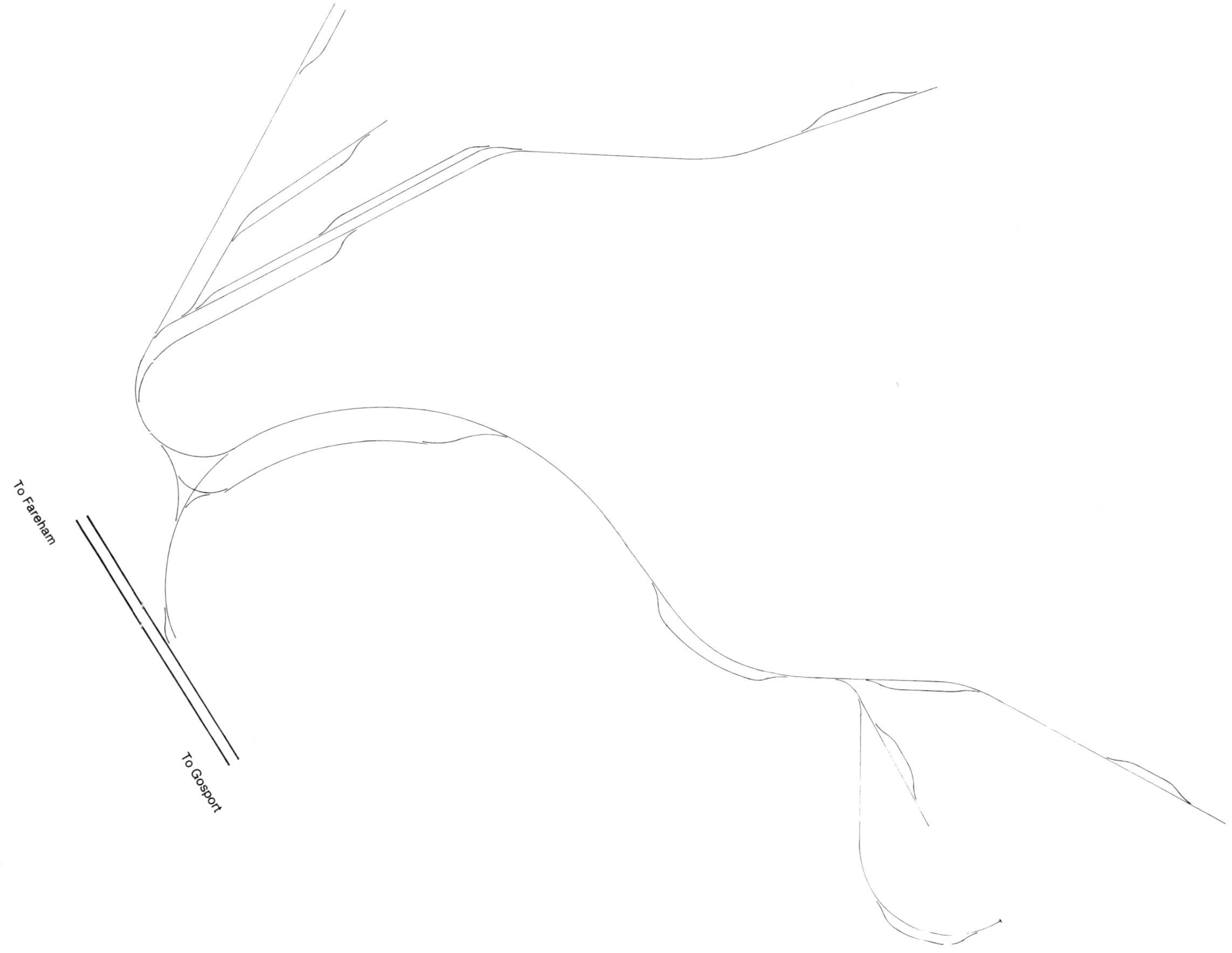

An unidentified through working just south of Fareham with the train running past open land near Redlands Lane. The lower quadrant lattice post signal and telegraph pole both typical features of the railway for many years. *National Railway Museum, Townley collection*

Looking north from Fort Brockhurst level crossing with the connection from the siding just visible. *D. Callender*

Nearing the end of its journey, an Alton – Gosport push-pull working leaves Fort Brockhurst and rejoins the single line. *L. Elsey*

Just before Brockhurst station on the right-hand side was another private siding and as was typical of the area, again for admiralty purposes. After this came the level crossing with Military Road, before the train would come shuffling to a halt at the station.

The main buildings here were on the 'up' side, the 'up' platform forming an island to also serve Lee-on-the-Solent trains. The area around the station was low lying ground, this prone to flooding under exceptional circumstances. Such was the case on two separate occasions in February 1925 when due to exceptional rainfall both lines north of the station were flooded to '....a considerable depth....', it being necessary to instigate a 15 mph speed limit to a point some 200 yards north of the level crossing, whilst the accumulation of water prohibited the safe working of the east crossover points.

Upon leaving the station the Lee-on-the-Solent line could be seen curving away to the right, although signs of this are soon lost. Sometime after 1910 there was a proposal for a 10-chain radius curve connecting the Lee line back towards Gosport. This would have formed a double junction to the main line linking in just north of Cambridge Road, with a single connection on the branch. Two new signalboxes would have been required to control the layout. But the advantages obtained are doubtful, there being little need for a triangle so close to the existing Stokes Bay junction. Perhaps for this reason nothing further is heard of the proposal.

The urban development of the area, coloquially spoken of as Camdentown, can now be seen on both sides, Ford Road being just one of the streets on the left whilst that to the right is of more recent building. This housing continues the short distance to Cambridge Road level crossing and the site of the fatal accident of 1895 already described. As built, the gatekeepers hut was on the 'up' side of the line, south of the crossing. A replacement structure was subsequently provided on the opposite side of both line and crossing. Two hand operated gates were provided either side of the road with an additional wicket gate for pedestrians.

The crossing is mentioned in the railway minute books several times over the years. The first on 29th November 1916, when at 8.58am the gates were run through by the 8.10am empty train from Eastleigh to Gosport. Gateman Alderman failed to hear the warning bells indicating the train's approach as he was at the time indisposed in the lavatory! The cost of repairs was put at £6.10s.

For some years, in the 1920s, the crossing was the place of work for one Walter Frederick Game. This man lost both legs in a

The Lee-on-the-Solent shuttle service at Fort Brockhurst.
R.C. Riley collection

CAMBRIDGE ROAD LEVEL CROSSING

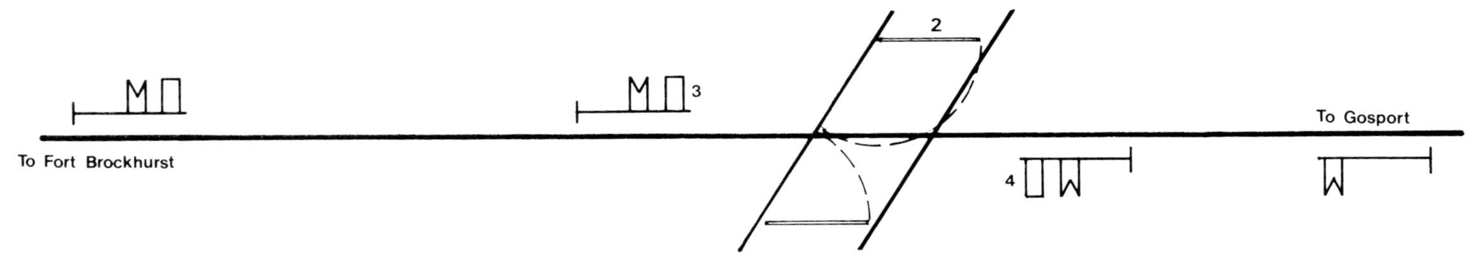

LEES LANE LEVEL CROSSING

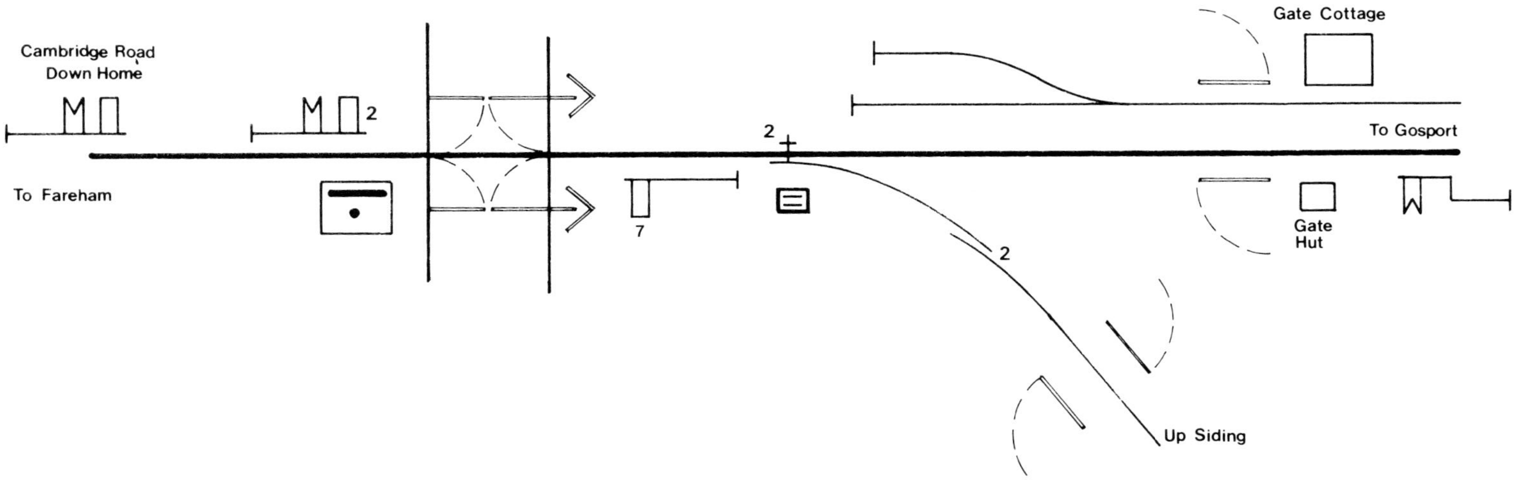

The later Cambridge Road Crossing signalbox and of basic SR design. *Sean Bolan*

shunting accident in 1903 after which the compassion of the railway company showed through by providing him with a special tricycle to use between his house and his then place of work on the Lee line. This item was reported as requiring repairs in 1914, the £3 9s 6d bill again paid by the South Western.

One may reasonably assume that Game was a well-liked employee, although his favours floundered somewhat on 22nd January 1921 when 'A12' 0-4-2 No. 655 ran through the closed gates in the course of running light between Fareham and Gosport. The railway's enquiry into the accident read as follows and provides a useful source of information on the workings of the period;

'.... the gates are worked by hand, there are no signals but there is a relay bell connected to the signalboxes on either side. At midday the 10.45am goods from Eastleigh stopped at the crossing according to usual practice to put out a bag of coal for the gate hut, as the crossing keeper on duty is maimed, having lost both legs and uses artifical limbs. One of the trainmen carries the coal to the coal house at the rear of the gate hut for him. The light engine approached as Game was later putting away the coal and he did not hear the whistle.'

Game was held responsible for the incident for failing to respond to both the relay bell and engine whistle. He was though only warned, '....as sympathy felt for his disability.' The cost of repairing the gates was put at £10. The comment concerning the lack of signals is interesting, although more likely the reference is to the box being a gate box rather than being a block post for certainly in later years some signals were provided.

Just a few years later, on 24th November 1924, the same level crossing was the scene of a similar incident. This was when,

Lees Lane Crossing signalbox, previously referred to as Forton Junction. *Sean Bolan*

at 9.20am, gateman Gould is reported to have opened the gates to allow a milk cart to pass on the road but then failed to close them for the 9.25am train from Gosport which was due a short while afterwards. Gould had by then retired to the signalbox for his meal and stated that he received no signals or bells to intimate the trains approach. The inevitable resulted and the gates were again run through. The bells were later tested and found to be in order. Gould was reprimanded and comment passed that he had only one arm due to an incident whilst on company service.

It would appear that these crossings were not the most renowned places for safe working, for Cambridge Road again comes to note on February 4th 1938. At 5.36pm the Alton-Gosport push-pull train ran through the gates whilst Gould was attending to his boiling kettle. His explanation was that he did not hear the relevant bell signals which were passed between Fort Brockhurst and Gosport. Both he and Driver Tossell were held to be at fault.

The final reference to an incident at the crossing comes shortly afterwards and was in many ways reminiscent of the first described. It occurred on 9th October 1939 when 'K10' 4-4-0 No. 329 ran through the gates in stormy conditions whilst working light, tender first, between Fareham and Gosport. At the time there was a cyclist passing over the crossing, the report of the incident stated, '....minimal damage to loco. Cyclist had cycle damaged....' The gateman again was answering a call of nature when the warning bells had rung.

It is interesting to note that both the latter two crossing keepers involved were men who were severely disabled, it is reasonable to assume that such posts were entrusted to such individuals.

The course of the line is now a gentle left-hand curve, continuing past further residential development either side of the line and under Anne's Hill Arch, an interesting structure, part arch and part span, consequent upon a widthways extension of the road above around 1914. The whole structure had a minimum 27 feet width at track level with a measurement of 35 feet 10 inches on the skew. It was also the penultimate bridge prior to the terminus. After this comes another level crossing at Lees Lane, believed to have been originally called 'Forton Junction', the name being changed about 1910 to 'Lees Lane Junction' and then finally 'Lees Lane Crossing'. The first two were in recognition of the time when it controlled the west curve leading to the Stokes Bay line.

The signalbox here was on the 'up' side of the line at the Brockhurst end of the crossing, the building of typical South Western design having a brick base with wooden superstructure under a tile roof. Inside was an 18-lever forward facing frame, shortened to 9 levers from 1934 onwards. As a matter of interest the four gates themselves are recorded as being worked by a wheel during single line days, whilst hand operation had been the norm in the days of double track and consequently heavier traffic.

Just south of the crossing was a pedestrian footbridge, locally known as Jacob's ladder due to its awkward approach steps. These had been improved in 1890 after a petition had been received from local residents, the L & SWR engineers agreed to changes to the bridge, steps and level crossing gates. The estimated cost of this was put at £280. From Lees Lane it is a short distance now to Gosport station with Carlyle Road running parallel to the line on the right. But first on the right is the site of the private siding for Ashley Wallpaper Co. whilst shortly afterwards comes the divergence of the Stokes Bay line, the actual junction name of 'Forton' taken from the nearby area name.

'Jacobs Ladder' footbridge and Lees Lane Crossing from the Gosport side, the sighting board behind the stop signal was to allow ease of visibility to approaching trains.
Pamlin Prints, Croydon

Shortly after this comes Moreland Road crossing. A footbridge was provided here in 1933 at a cost of £266 thus saving £207 p.a. in wages. It was in reality a simple occupation crossing leading to Toronto Place, for some time the solitary housing within the rail locked area of the Gosport triangle. To control the crossing there was a small gatekeepers hut on the south side and a house provided for the crossing keeper on the opposite side of the line. After this, again on the left, comes the open greenery of Forton Drill Field and Recreation Ground following which Gosport Junction is passed to the right, the site of the other side of the triangle and original access to the Stokes Bay line. Immediately afterwards comes the pedestrian footbridge at Queens Road, before the train passes the engine shed and goods yard to enter the platform at Gosport station.

Above: Another local goods and this time with '700' class No.30350 heading south near Gosport. *L. Elsey*

Right: Viewed from the terminus at Gosport with the signalbox and engine shed shown. The former 'down' line is relegated to siding status and now affords access to the goods yard. *R.C. Riley*

The station around 1910 with original roof and various contemporary enamel advertisements. Supporting the roof from the platform are cast iron pillars. These replaced masonary supports which were susceptible to being struck by carriage doors.
National Railway Museum

The Line to Clarence Yard

East of the station, a single line of rails continued on towards Clarence Yard. In earliest times this diverged off the goods line but was later altered to run from the middle line. There were then two level crossings, the first over Spring Garden Lane and the second Mumby Road, this latter a busy thoroughfare linking the Portsmouth ferry with Brockhurst and Fareham. The two level crossings were separated by a small triangle of grass . The gates themselves were hand-operated and worked as necessary by men from the Clarence Yard. The line then curved slightly left and ran in a shallow cutting under an arch in the ramparts of the old town. From thence its course was across Weevil Lane before dividing to either the main yard and former Royal station or continuing on via embankments and a bridge to a fuelling jetty alongside Weevil Lake.

Above: A transfer working into Clarence Yard crossing Spring Garden Lane. The gates here allowed for the railway to be completely protected from the road but it was practice to only close one as shown here.
R.C. Riley

Left: The rural appearance of the line crossing Spring Garden Lane with No.30175 in charge on 27th June 1950.
D. Cullum

Stokes Bay Line

Leaving Gosport station the line to Stokes Bay diverged via the east side of the triangle, this being a 15-chain radius curve and taking the line almost due south. It was a short distance now to Stokes Bay Junction, the point where the connection from the other side of the triangle merged. Today the path of the curve is still marked by the gardens of the houses in King's Road, whilst the area within the triangle was for many years allotments. One of these was cultivated by a man who specialised in carnations. A driver once when turning his engine at the dead of night helped himself to some of the prize blooms, the loss being discovered and the driver sacked! At the end of the curve, the line on the other side of the triangle, from Lees Lane Junction, joins with Stokes Bay Junction signalbox on the right-hand side of the junction. This, in appearance, a typical South Western structure and similar to Woolston signalbox on the Netley line.

The double track continued south and under Love Lane footbridge, having a single span 48 feet 4 inches wide across the railway and some 89 miles 21 chains from Waterloo. The actual bridge number was 37A, the suffix indicating it was not an original structure but added sometime after the line opened. Continuing almost straight, the formation entered a shallow cutting, culminating in the twin brick arches of Bury Road bridge before entering the small station at Gosport Road. The main buildings here, if one can call them that, were on the 'up', or

Above: The junction with the Stokes Bay line just outside Gosport station. The bracket signal previously carried an arm for trains destined for Gosport Road.
D. Cullum

Right: Looking back from Gosport junction towards the triangle and main line. The former junction box stood on the left side of the line although at the time this view was taken in 1950, the triangle was used solely for engine turning with the points in the foreground hand operated. *D. Cullum*

Gosport Road looking north in the early 1930s and with the twin arches of Bury Road bridge at the end of the platform. *D. Thompson*

Fareham side both platforms linked by an iron trellis footbridge which also provided for an accessway across the railway from the direction of Linden Grove.

Leaving the station the cutting gave way to a shallow embankment and destined to continue almost unabated the whole of the short distance to Stokes Bay itself. To the left Cleveland Road ran parallel to the line whilst on the right the back gardens of houses in Linden Grove came up to the boundary fence.

Just a few yards further on came Workhouse Bridge, locally referred to as Workhouse Viaduct, the name as might be expected taken from the old Alverstoke House of Industry or Parish Workhouse which had once stood near the spot. Unfortunately details of the original viaducts here and at Haslar have not survived, although these would appear to have been substantially longer than those of later years. Plans have been found showing both these and the Haslar viaduct as being reconstructed just after the turn of the century, the accompanying notes stating the original viaducts to be '....superseded by embankments and wrought iron bridges....'. At the same time some of the under bridges were similarly dealt with. The timing for the rebuilding is interesting as it was at the same period the new ballast siding was brought into use a little further south. Possibly the reconstructed bridges would allow for heavy trains in conjunction with the working of the new siding.

In connection with the reconstruction, temporary single line working was instigated between Gosport Road and Anglesea Crossing, both signalboxes being upgraded to block posts and equipped with tablet instruments. As would be expected, the Railway Inspectorate were informed of the change, noting the length of line affected as being some 1,200 yards long. It was anticipated the work would take six months to complete, the work involving replacement of the existing cast iron girders and timber supporting piles with embankments and wrought iron bridges. Workhouse 'Creek' bridge, as it was described in the plans, was dealt with first by taking possession of the 'down' line to begin with and allowing for the sinking of 4 feet 6 inches diameter supporting piers to a depth of 40 feet. After this the 'up' line was dealt with and then Haslar bridge attended to in similar fashion. A little later, in 1904, it was reported that restrictions had been withdrawn for heavy engines on the line.

For a matter of a few hundred yards the line continued its course on an embankment, swinging slowly right in the process and running part way parallel with Mount Pleasant Road. Then the second of the viaducts was reached, this one, over Haslar Creek and after rebuilding having one span of 56 feet, slightly longer than the previous. Both bridges were reconstructed with three girders, one being central between the two lines.

This bridge came to be referred to as 'Jackie Spencer's bridge, apparently so called after the railway employee who had at one time had the task of patrolling the immediate vicinity from a small wooden hut erected nearby. It was reputedly the habit of some local residents to steal lumps of chalk from the railway embankment which were then used for domestic cleaning purposes, the practice having become so widespread that the very stability of the embankment was threatened.

From Jackie Spencer's bridge the line again swung right, to pass under the twin arches of Clayhall Bridge. This consisted of twin arches, each of 13 feet 3 inches clearance and identical in shape and design to the earlier Bury Road arches.

As described in the first chapter, the area around here had been worked from early times for both brickmaking and gravel, there being signs of an old gravel pit to the right of the line. On the left though, the South Western found sizeable offerings of

WORKHOUSE CREEK VIADUCT

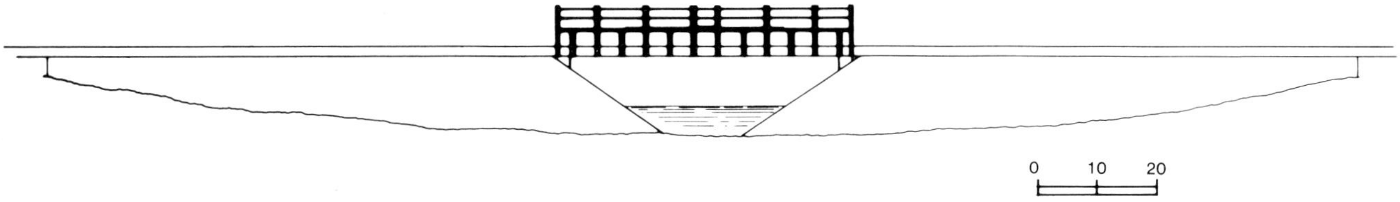

ANGLESEA LEVEL CROSSING SIGNALLING DIAGRAM

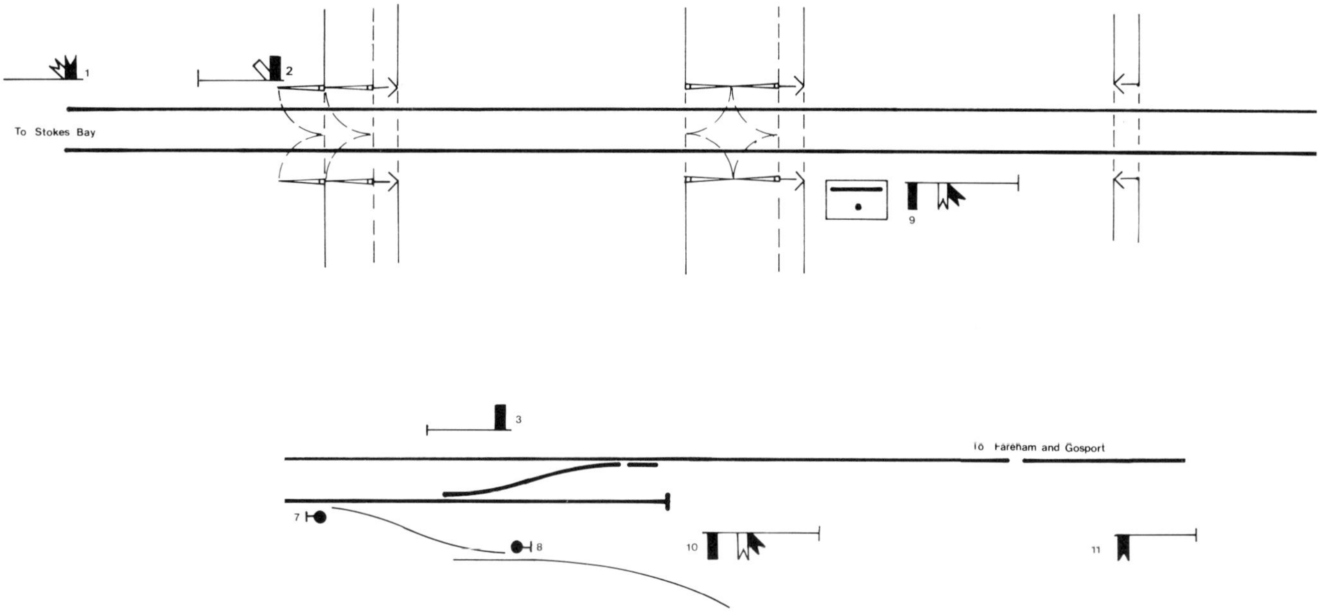

Two views of the bridges south of Gosport Road which carried the railway over Workhouse Creek and Haslar Creek. *Both D. Callender*

hardcore and ballast. It was decided around 1900 to excavate this and so provide material primarily for use in connection with the Meon Valley line then under construction between Alton and Fareham. Accordingly Board of Trade approval for the new work was first sought on 10th September 1901, the intention being to form a trailing connection into the 'down' line just north of a pedestrian crossing on the railway known as St. Mark's crossing. The connection would be controlled from the existing Anglesea Crossing box. This was to be converted from a crossing box to a 'lock post' at an estimated cost of £155 and chargeable to the Meon Valley construction account.

The new connection was inspected by Major Pringle of the Railway Inspectorate on 30th January 1901. He made the comment that the siding was only for 'temporary use', for if permanent, alterations would have to be made due to the close proximity of St. Mark's crossing. Having only a trailing connection into the main line and no crossovers known of on the Stokes Bay at all, it is interesting to speculate on the method of working full wagons from the site. Plans of the area show a single line of rails to standard gauge running back into the quarry and referred to as a tramway. Was it then a question of propelling wrong line back to Gosport? Probably unlikely, more feasible perhaps is that the existing single line connections in conjunction with the rebuilding of the viaducts were retained for the purpose.

As mentioned, the controlling signalbox was at Anglesea level crossing and situated to the east of the line on the Gosport Road side of the crossing. In typical railway fashion the name Crescent Road crossing would probably have been more accurate. No structural plans or photographs of it appear to have survived, although from odd snippits of ground elevations found it is known to have had outside steps and possibly therefore of a similar type to Stokes Bay junction. Inside was a 12-lever frame, extended to 14 levers for the time single line working was in force.

The gradients on the line were ranged in a slight 'switch-back' manner, the final approach to the terminus on a climb of 1 in 290 followed by a section of level ground onto the actual pier. But first the line had to pass another level crossing, this time St. Mark's Road and having four gates shown on plans as opening in a somewhat unusual fashion being controlled from a small gatemans box. (It is not clear if this was a standard design signal-box classified as a gatemans hut or a ground level structure proper). Alongside, a small railway owned cottage was home for the man in charge.

The embankments then took the line over several small culverts, one the former moat round the southern forts of the area. With the ground then falling away the line passed both the high and low water marks on the shoreline to finally surmount Stokes Bay Pier itself, not quite 91 miles from the London terminus. Stokes Bay Pier was a formation of girders and spans numbering fourteen in all. All but one of these had 28 feet 3 inches clearance with the last 9 feet longer. Unusually there were two platform faces to one track, both partially covered by a timber roof. The 'down' line connected into the platform for the use of passenger trains. A small signalbox, again of unknown type and located slightly north-east of the terminus, controlled the layout.

Early accounts tend to portray the Stokes Bay line as suffering badly at the hands of vandals. One example in 1911 when Signalman Warwick at the terminus apprehended a boy who had just thrown a bucket at a passing train. The villain received six strokes of the birch as sentence from the local court.

The approach to Stokes Bay terminus shown in its final years. From here the Isle of Wight was but a short distance. *Sean Bolan*

Stokes Bay pier station as it probably appeared around the turn of the century. *From a drawing by Ivan Bovey*

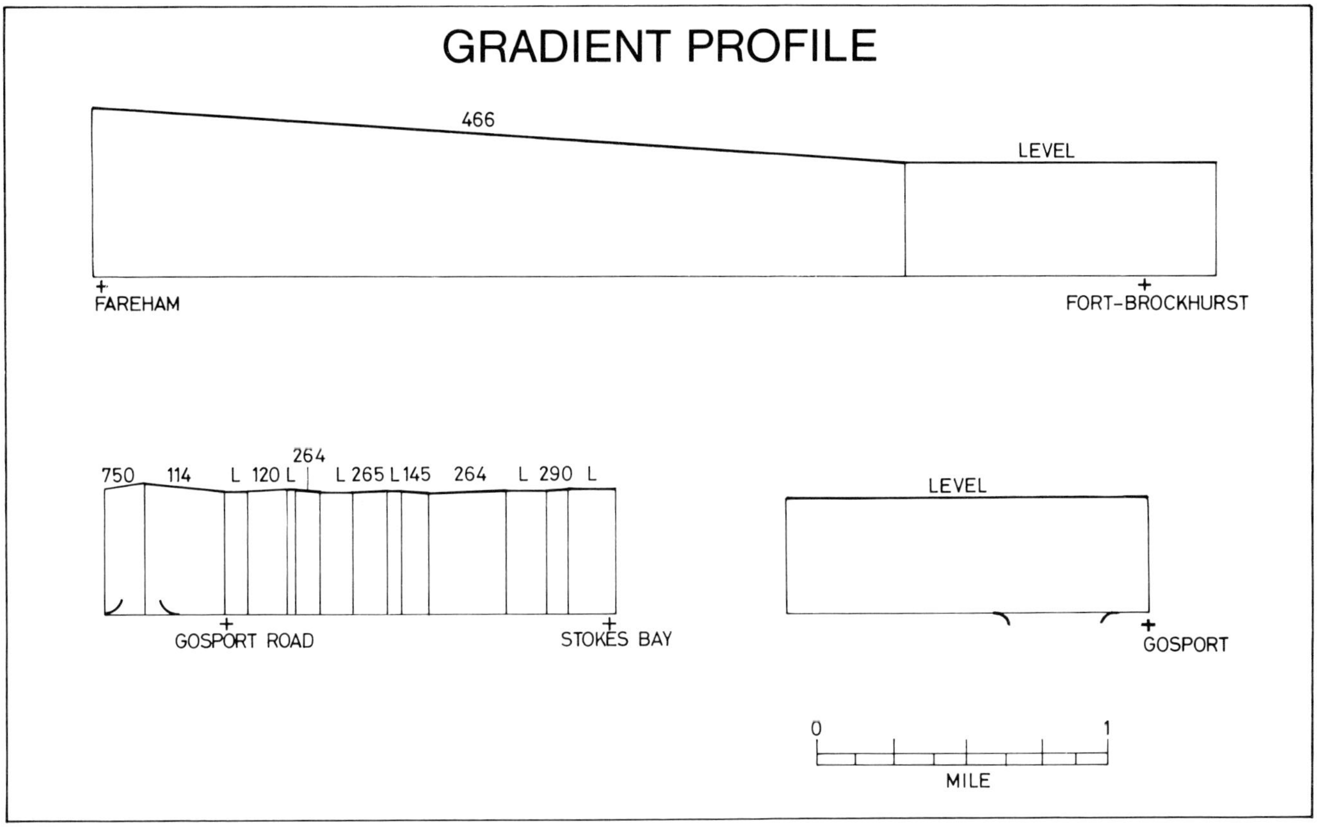

Lee-on-the-Solent Line

Commencing from the rear platform at Fort Brockhurst station the single track Lee-on-the-Solent branch curved sharply away southwards so that before very long it was running at almost right angles to the main line. As would be expected the railway was hardly suitable for speed, whilst in addition, the closeness of the various halts and the not always perfect condition of the permanent way, precluded any attempt at fast running. Not unexpectedly then several accounts have emerged of minor derailments and mishaps on the line, some of these comical to the extreme.

To set the scene one may perhaps first describe an incident which took place on 17th July 1920 at Fort Brockhurst. The 12.50pm departure for Lee left the station only to run straight into a set of stopblocks demolishing these and totally derailing not only the engine but the front bogie of the first coach as well. It transpired that shunting had been taking place prior to the train leaving and the staff failed to ensure that the points were re-set for the passenger train. As might be expected the driver, fireman, porter and guard were deemed responsible and 'dealt with'. It was fortunate that there were no injuries amongst the 20 passengers reported to be on board and perhaps the most amazing thing is that no complaints were made by the travelling public concerning the delay to the train or subsequent missed connections on the return journey. A replacement engine and coach were summoned from Gosport and the derailed vehicles attended to by the Eastleigh Breakdown Gang.

Leaving behind for a moment the misfortunes of operations on 'the branch' as it was referred to by the Gosport men, the little line of railway continued to curve almost to an 'S' shape, so that before very long it was running parallel to Military Road with open fields to the left and what is now HMS Sultan, the Royal Naval engineering school on the right. It was near here that trains from Lee would sometimes make an unofficial stop, allowing schoolchildren destined for the nearby Garrison School to alight near to where Baden Powell Road was later sited. There was then a small gated level crossing at Pound Lane. After this the line started to curve to the right as the train began to slow from its already leisurely pace to pass another level crossing and immediately after stop at what was for many years Fort Gomer Halt.

The level crossing at the halt was the scene of several accidents involving trains and vehicles, one of these at about 6.19pm on 26th May 1924, when a Lt. Commander Richardson failed in his motor to beat the train. His car was crushed against the railway causing some injury to him. The train concerned, the 6.15pm from Brockhurst to Lee, was delayed by some 26 minutes. Two years later, in 1926, the brakes of a motorcycle failed just as a train was approaching from Lee. Both rider and pillion passenger were fortunate to wake up in nearby Haslar Hospital. Then at 1.34pm on 14th September 1934 a car driven by a well known railway photographer Mr. O.J. Morris ran into a four-wagon freight train in the course of crossing the road. This time both car and train were damaged, the official report giving no information as to the state of the driver. Finally, at 2.25pm on 19th July 1935, as the 2.20pm goods from Lee to Brockhurst was at the crossing in the charge of 'Terrier' No. 2234, there was a collision with an RAF lorry. The locomotive mounted the front of the road vehicle derailing itself in the process. No blame was attached to the engine driver and no-one was injured.

Fort Brockhurst and junction with the Lee line, the connection running off to the left with the platform for Lee trains hidden by the trees by the main station building. *R.C. Riley*

Leaving Fort Gomer Halt the curve ceased and the line passed over the site of a small foot crossing abolished as long ago as 1910. To the left there was open heathland. Up to now the route has almost been at ground level since Brockhurst but this began to change to a shallow embankment culminating in a small stream underbridge carrying the first of two courses of the River Alver. The bridge itself is 1 mile 23 chains from the junction and had an 11 feet 6 inch span with a height of 6 feet 6 inches. Some 250 yards further is a second bridge of identical size and design. The method of construction used was four steel girders on a decking of old rails laid on wooden stringers and supported on piles with concrete abutments. These were the only two underbridges on the whole of the line.

Just afterwards there was a small footbridge, the railway regaining its gentle right-hand curve before entering a shallow cutting to pass Browndown Military Camp on the left. Its course then took it parallel to another, Military Road, being unmade at the time the railway was operating but now well known as Portsmouth Road, running from present-day Browndown roundabout at Gosport towards Lee.

A slight left-hand curve brought the railway to the appropriately named Browndown Halt and preceeded by yet another level crossing taking Military Road at right angles across the line. Surprisingly, in view of the number of incidents recorded elsewhere, no accidents are reported here. The platform at Browndown was in-keeping with Fort Gomer and on the 'down' side. Opposite were later to be the first signs of the residential development of Lee-on-the-Solent at Portsmouth Road and Chester Crescent, whilst behind were fields followed by a short drop to an area known as Browndown Ranges. These continued southwards for a matter of a few hundred yards to the shore and as the name implies were used for military training.

Within these ranges there existed a 2 feet 6 inch tramway. This ran in three sections from a central repairing shop and store to one of three targets almost on the coast line itself. A fourth tramway to a new aiming point was added by 1931, by which time also slight modifications had been made to the course of one of the lines. The maximum length was some 800 yards from the repair shop to the target shown as No. 1 on the plan. Little is known of the origins or operation of the tramway, although it appears on maps of 1898 and was known to be 2 feet 6 inch gauge and would probably have been hand worked.

North of the railway a gravel pit is also marked on the OS maps, but it is not clear if this ever had any significance to the railway.

The railway then ran straight and level with Portsmouth Road with a panoramic view of Southampton Water and the Isle of Wight on the left. This continued for almost half a mile until swinging left slightly and then immediately right on a curve of 10 chains radius to descend through a shallow cutting on a 1 in 66 gradient and still with the public road almost parallel. The route had also brought the railway almost onto the sea shore. A halt was opened in the vicinity called Elmore, the name taken from nearby Elmore Farm and allowed passengers immediate access to the beach near to the present day sailing club.

It was just near Elmore Halt on 9th September 1921 that there was a spectacular accident. '02' 0-4-4T No. 177 was hauling the two coaches of the 5.25pm passenger train from Lee to Brockhurst when the penultimate pair of wheels on the rear coach left the rails and ran a distance of 280 yards on the sleepers before the train could be brought to a stand. The coach was detached and the train continued to its destination although two later trains were cancelled until the Eastleigh gang arrived to deal with the situation. It took them just 100 minutes to re-rail the vehicle. The cause was never established, although it was suspected to be the condition of the track, described as '....appreciably worn....'.

With Elmore Halt passed the railway continued on a shallow embankment between the shore line and Marine Parade East the latter roadway running almost the whole length of the front at Lee and a continuation of Portsmouth Road. Road and rail here were separated by a narrow grass strip. There then followed a small ungated crossing as the single line swung right and then left to terminate alongside both sea and shops at Lee-on-the-Solent station with the renowned pier but a few yards distant.

A well-known copy of an old postcard showing the terminus. The use of three carriages possibly indicating the summer service.
Lens of Sutton

A superb study of Fareham station with a special working from the Bishop's Waltham branch. An 'M7' is sandwiched between the two 2-coach auto sets.
W. Gilburt

Chapter Seven

The Stations

The station at Fareham dates from the time of opening of the Gosport branch proper, having been first opened for traffic in 1841 and then like Gosport, closed almost immediately afterwards due to the state of the nearby tunnels.

At the time the railway first reached the town, the buildings were of a different style from that more recently recalled, the original construction at a cost of only £1391. Unfortunately full details of the original facilities are not available, although it is known that from 1841 to 1889 there were only two platforms, probably connected by means of a wooden board crossing at track level. The main station buildings were also on a different site from that used later. These were in some way reminiscent of the style utilised at Bishopstoke and elsewhere and thus maybe owe their origins to the designs of Tite. On the 'down' side the goods shed was at right angles to the line and in its approaches were two wagon turntables, one on each running line. Public access to the passenger station was in a similar location to that existing later on. No signalboxes are shown on early plans.

As would be expected, over the years several changes affected the locality, and space prohibits a full account of the history of this fascinating station. However, some changes are detailed. The first of these was on 1st September 1848 when south of the station a junction was formed taking the line eastwards towards Cosham and then Portsmouth. The proximity

The approach to the cab yard and buildings on the 'down' side.
SR Magazine

Fareham

Busy times at Fareham with trains to and from both Portsmouth and Gosport in view. *D. Callender*

An unidentified 'T9' rounding the curve into the station with what would appear to be a through Portsmouth – Bristol service. The severity of the curve is clearly seen in this photograph. *D. Callender*

No.726 at Fareham with the 7.45 pm Gosport – Alton service in July 1947, the cattle pens are just visible on the extreme right. *J.H. Aston*

Another early postcard view of Fareham and clearly showing the priority to Gosport trains. *A. Harris collection*

of the various water courses adjacent to the railway caused the new line to take a sharp curve towards Cosham and for many years was a difficulty in through running.

The new line brought with it a considerable increase in traffic, much of it to and from the west country and routed via Bishopstoke and Salisbury. Then, on 1st July 1889, the line from Netley was extended through to Fareham connecting in with a junction north of the platforms, again with a corresponding increase in traffic and now making it possible to reach Southampton with ease.

View north from Fareham with the various signals reading for the Netley line, deviation and tunnel routes. *L&GRP courtesy David & Charles*

FAREHAM c.1860

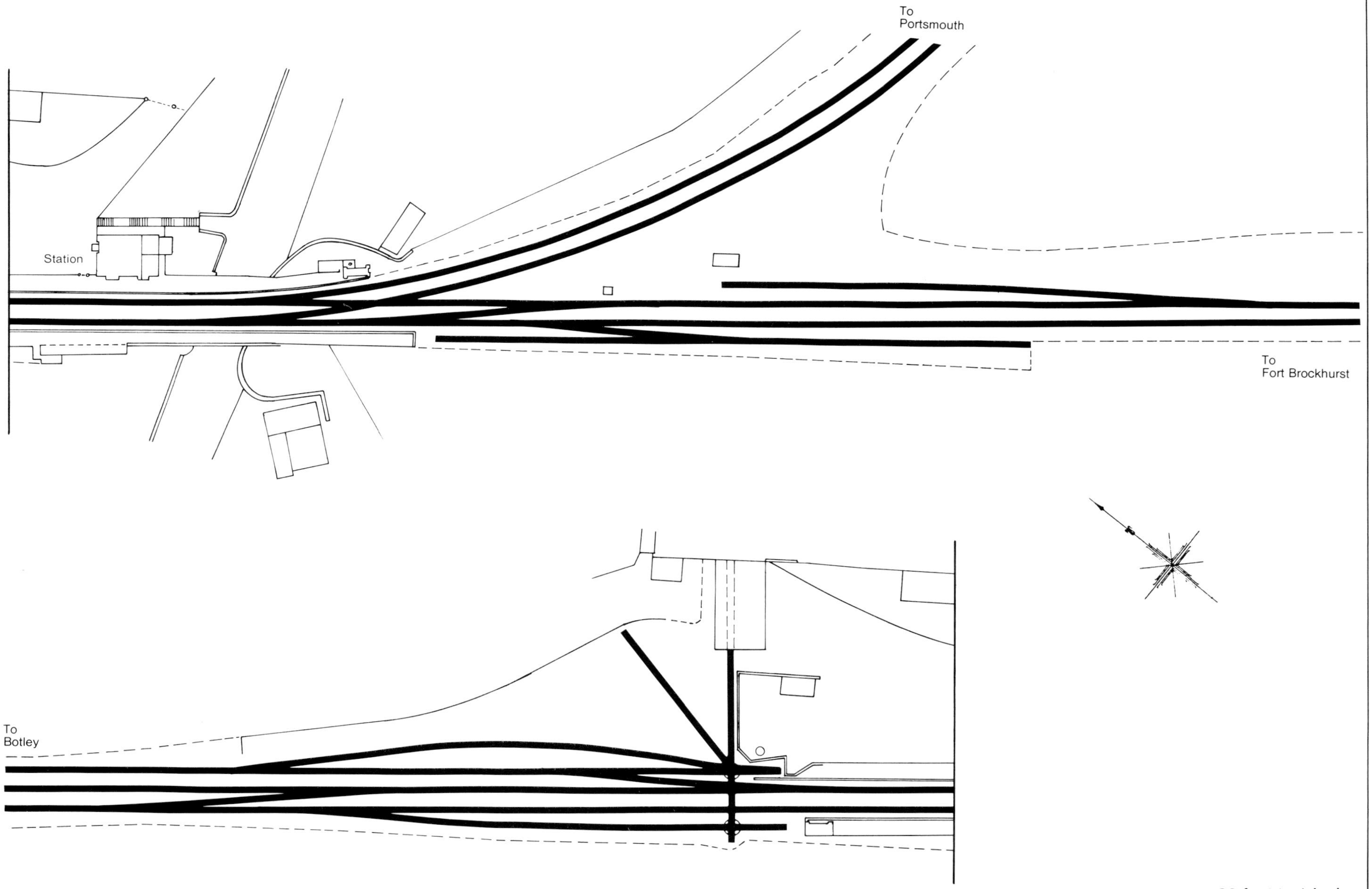

80 feet to 1 inch

FAREHAM 1898

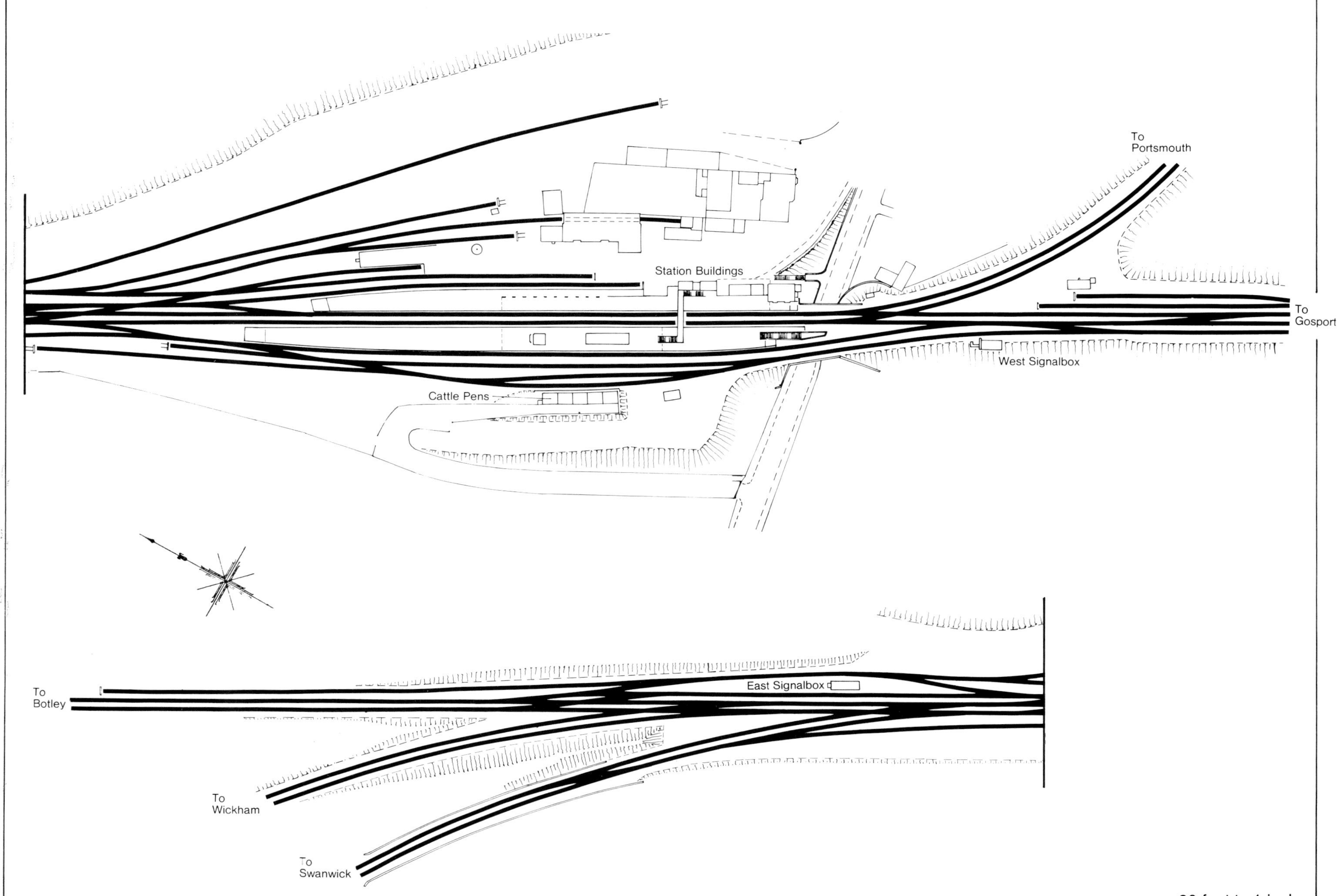

The north end of the station with the 2.50 pm Waterloo service, via the Meon Valley, awaiting departure. *SR Magazine*

At the same time plans were prepared for a complete rebuild and redevelopment of the site. The existing 'down' side buildings were replaced by a new single-storey building fronting the 'down' line platform face. This was used by trains from either Netley or Bishopstoke lines and destined for Gosport or Portsmouth. Opposite was an island platform, numbered 3 & 4. Platform 3 could be used by all 'up' services, although only trains from Gosport could reach Platform 4. Wooden and glass canopies provided a degree of passenger protection for much of the platform length, whilst the two platforms themselves were connected via a substantial covered footbridge. The new buildings were supported on a series of brick arches sunk into the ground.

Opposite Platform 4 various sidings fronted some sizeable cattle pens, whilst on the approach side of the station was a new brick-built goods shed which had replaced the previous structure. Adjacent to this were numerous mileage sidings as well as a dock at the Botley end of the 'down' platform, the wagon turntables being removed in consequence of the rebuilding.

Passenger access to the site was via one of two points. The first for pedestrians was via platforms 3/4 and involved the use of a flight of steps to the main Southampton-Fareham road. This came out between the two bridges carrying the railway over the road and is clearly shown in the photograph on this page. These steps were provided from 1889 when the additional arch was erected over the main road. Ticket security was provided by a small wooden gateway on the island platform itself. The second method, which was also the only means available to cabs and private carriages, took the form of a sweeping roadway from West Street, Fareham, past the Railway Hotel on the right to finish up alongside the main buildings. A continuation of this route took

View from the Gosport line showing the route from Portsmouth coming in from the right. *SR Magazine*

Looking down the steps between the road arches accessible from the island platform. *Sean Bolan*

FAREHAM STATION BUILDINGS 1889

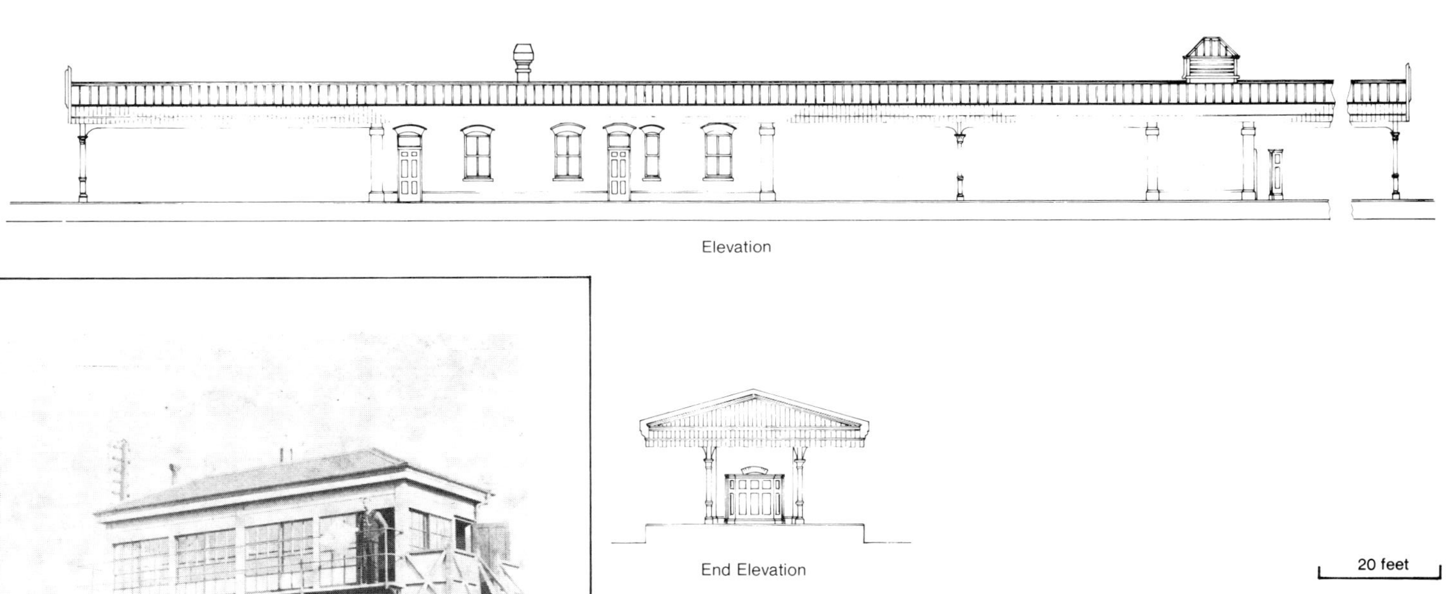

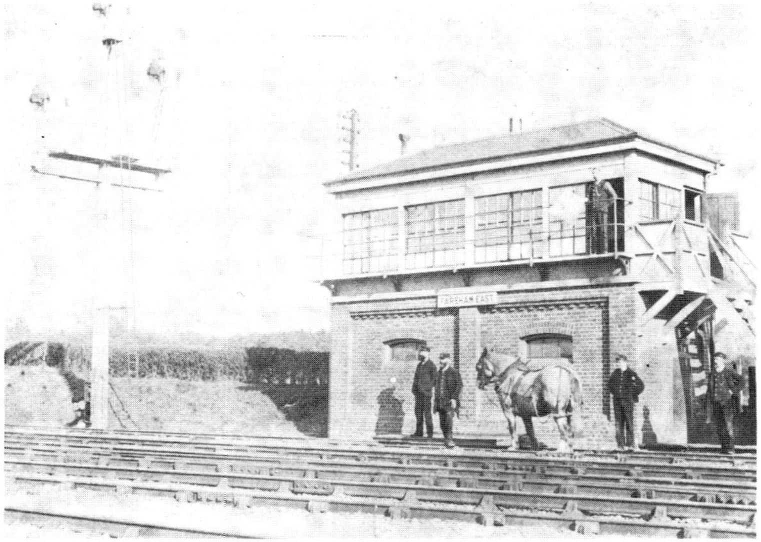

Above: Fareham East signalbox with staff and shunting horse posing in front. *Dennis Tillman*

Right and far right: The lineman's cabin at Fareham. The S&T lineman was responsible for the electrical side of the signalling on the Gosport branch. *British Rail* vehicles directly to the goods shed and yard. Additionally, road access was available to the cattle pens from the main road.

To control the layout, from 1889 onwards, two signalboxes were provided. Respectively Fareham East and West and both of similar style.

The final stages of development came from 1904 onwards, when north of the station, back towards Botley, a double track deviation round the troublesome tunnels was built. Ironically, in an effort to rid themselves of the tunnel nuisance, the engineers created another, for slipping of the cutting sides on the deviation was to continue throughout its life. The original route was then mainly used for Meon Valley trains, although various changes to the track layout at Knowle Junction, north of the tunnel, altered this several times over the years. Then in 1938 a private siding was established just off the Netley line to serve an Air Ministry depot.

The rebuilding of the site had been necessary due to a continual increase in traffic. The area of Fareham was renowned for market produce, wagon loads of which were handled daily. In addition the yard took on the role as one of the main marshalling and dispatching points for soft fruit traffic in the eight weeks or so this was in season. Whole trainloads of strawberries and the like were handled en-route to the London markets. There was also a respectable livestock market in the town causing trainloads of cattle wagons to be dealt with as required.

On the passenger front there was for many years local services to Portsmouth, Gosport, Southampton, Eastleigh and Alton, these last named services often starting from a bay at the

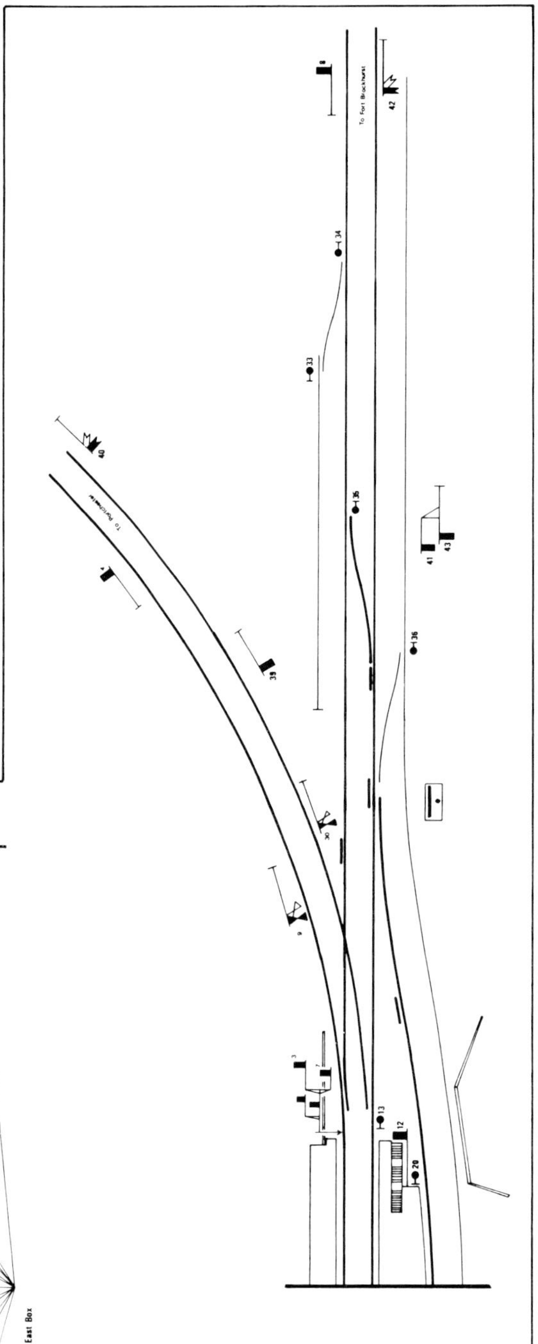
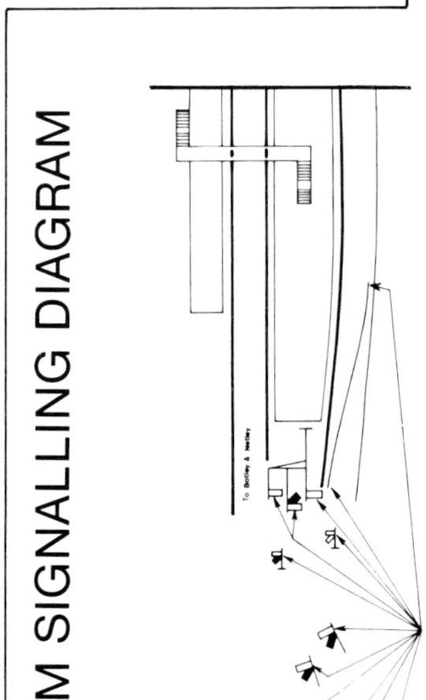

Above: Ex-works '700' No.30690, running-in on the morning goods from Eastleigh to Gosport in 1959. *David Fereday Glenn*

Right: Handing over the tablet at Fareham West. *D. Callender* tunnel end of the 'down' platform, referred to as Platform 1. In addition there were through trains to Brighton and Cardiff. Some of these attaching and detaching vehicles to and from the Southampton line. Understandably there was a valued commuter traffic from very early on, some of this schoolchildren to one of the four senior schools at one time existing within the borough.

Despite an almost continuous series of movements throughout the day, no locomotive shed or full time pilot was provided, motive power requirements being catered for either from Gosport or Eastleigh according to need. It was the practice to utilise the locomotive of the Fareham – Gosport, or Fareham – Alton shuttle services for shunting etc. as appropriate. The timetable also allowed for various light engine movements to cover the various needs.

As an example, in the spring of 1924, 53 passenger and 10 goods trains were dealt with in a typical 24 hour weekday period in addition to 5 light engine workings. It was possible for many years to purchase weekend return tickets from Fareham to London travelling either via Eastleigh or the Meon Valley routes.

Interestingly, despite the importance of the station, there was no local goods agents for the town, the nearest listed being Messrs. Chaplin & Co. at Gosport, although it is likely suitable arrangements were made to cover what was a widespread area.

With the closing of both the Gosport and Meon Valley lines in recent times, Fareham has changed to a style more akin to a through station rather than an important junction, although it is still possible to travel on either the Netley, Botley or Portsmouth lines. There have been changes also to attempt to increase the speed limit at the west end, with the Portsmouth curve now relaid on a wider sweep involving the building of a new bridge over the A27 to relieve what had become a recognised road bottleneck. This has meant that the former platforms 2 and 4 are now the respective 'up' and 'down' lines with Platform 3 a 'stub' used for the Fareham – Eastleigh shuttle. In the goods yard too there has been a change, the daily pick-up goods and consequent shunting now cast aside to memories. In its place now are block workings of aggregate from the Mendips, the whole layout being controlled from the Eastleigh Signalling Centre.

Fareham 'down' side platform buildings showing a good deal of parcels business. *Sean Bolan*

A none too common sight with 'D1' No.31739 on a Salisbury – Portsmouth Harbour working. *W. Gilburt*

Fort Brockhurst

The station here was opened in 1865 and became a junction in May 1894 consequent with the opening of the line to Lee-on-the-Solent. The station name itself changed from Brockhurst to Fort Brockhurst following approval by the South Western Traffic Committee meeting of 17th November 1893, and was due to confusion arising with passengers luggage being directed in error to Brockenhurst and vice versa.

When opened there were 'up' and 'down' platforms. A small brick-built stationmasters house surmounting the 'up' platform adjoining which were the station offices and basic facilities for passengers. The opposite platform had but a solitary wooden waiting shelter with porters room adjoining, access between the two platforms was across the track itself at the level crossing of Military Road immediately at the Fareham end of the station.

The military and naval connections of the area are prevalent from the earliest times, several of the adjacent road names having 'battle' or other historical significance whilst close by the railway was the garrison church. It follows then that military traffic was to be in the forefront of traffic handled and was recognised only ten years after opening with the provision of a siding for the admiralty just north of the station and trailing into the 'up' line. Surprisingly, before this there is no record of any sidings whatsoever at the station.

The new siding was laid in by the South Western themselves on a site previously known as Soliers Gardens. Inspection for the Board of Trade took place by Captain Tyler on 16th April 1875, his report is as follows;

> 'I have the honour to report for the information of the Board of Trade that in compliance with your instructions I have inspected the new siding connection at Brockhurst near Gosport on the London & South Western Railway.
>
> Safety points have been provided and are interlocked as well as the siding points with the 'up' line signals. But this interlocking is effected by means of wires connected with a lever in the signal cabin. It is necessary that the adjustment of these wires should in maintenance be carefully attended to.
>
> Subject to this observation I am of opinion that the Board of Trade may properly sanction the use of the siding.'

LSWR 2-4-0 No. 21 *Scott* at the Lee platform at Brockhurst on 26th December 1900. This engine together with No. 392 on page 29 were the regular performers on the branch for some years, whilst the coaching stock is probably that owned by the Lee-on-the-Solent Company.
Pamlin Prints, Croydon

As a matter of interest this is the first mention of a signal cabin at the station and it is likely that the wooden structure recalled as existing in later years is the building referred to above.

To facilitate shunting arrangements to the 'military siding', as it was termed (this despite the reference to it as an admiralty siding originally), a crossover was provided in 1894 between the main lines at the Fareham end of the station, whilst later in 1897 the siding itself was extended and believed to have been the time when an overhead gantry crane was also erected. It now ran parallel with a loading platform upon which there existed a baggage shed and toilets. Plans have also been found showing proposals for a full length troop platform around the same time and near the existing works, these though were not carried out. Gas lighting was, however, added in 1903. Privacy of work was reflected in a gate across the siding near to the main line, the siding referred to by the staff as 'the yard'.

As intimated above, much of the traffic handled was of naval or military origin, thousands of men carried in conjunction with a general upsurge in traffic around the time of the Boer War. Slightly later, the military fort from which the station took its name was used as a discharge depot. Special trains brought the

Looking north through Brockhurst in British Railways days and with the level crossing gates at Military Road open to road traffic. The 'up' side canopy is beginning to show its age by sagging in the centre. *Lens of Sutton*

'M7' No. 30479 on an afternoon Fareham service at the station in June 1950. The sleepers dumped on the platform opposite indicate that some relaying may have recently been undertaken. *D. Cullum*

thousands of soldiers back who would shortly afterwards be trying to find their way into 'civvy street'. Soldiers of different kinds were also handled, that of prisoners of World War 1, who arrived at the station en-route for the Lees Lane military prison.

The opening of the line to Lee changed the station to that of a junction, an arrangement being reached between the South Western and then independent Lee-on-the-Solent Company for a separate platform for the latter's trains. This was situated behind the existing 'up' platform and provided basic facilities in the form of a wooden platform connected to the main 'up' side by a gate. No shelter was supplied for Lee line passengers, although of course such protection was available at the main station.

Physical connection between the two railways was by means of a siding connection. This meant that a shunt move was necessary to place or remove a vehicle or train onto the light railway. Consequently no through trains as such could be run. The occasional through service that was provided required passengers to alight whilst the vehicles were shunted before rejoining them for the remainder of the journey. A similar situation existed in the reverse direction. The Lee-on-the-Solent service was little better than a shuttle and as it was a light railway no signals were provided. Operation was by means of a solitary train staff onto which a key was attached to unlock the various points as required. To develop the line direct access would have been needed, allied to changes in the operating pattern. These were never to materialise.

Being little more than a wayside station, few changes are recorded as having been carried out to the passenger and parcels facilities over the years, although mention is made of the provision of a small goods store for £50 in late November 1894. Another was a reported extension of the platforms. This was authorised on 20th July 1898 at an estimated cost of £208, whilst previous to this, in 1891, authorisation was given for the booking office to be altered, the platform extended on the 'up' side and a separate urinal built. The cost of all of which was put at £160. Some doubt exists though as to what was actually completed.

A closer view of the 'down' side. The cut-away under the platform would have been used for point rodding compensators and cranks, however, there is no evidence to suggest that a signalbox was ever placed on this side. *R.C. Riley*

Now in a very tatty condition, the signalbox at Fort Brockhurst stood at the Fareham end of the 'up' platform and contained a Stevens ground level frame of 19 levers. *Sean Bolan*

An afternoon Fareham-Gosport service entering the station in the summer of 1950. The level crossing here consisted of four wooden gates operated by a wheel in the signalbox. *D. Cullum*

Above: Just north of the station was the Admiralty siding and platform, although little used in its latter years. Access to this was via a gateway from nearby Military Road. *D. Callender*

Left: The unusual 'up' starting signal at Brockhurst with the post set at an angle to the signal arm. *D. Callender*

FORT BROCKHURST SIGNALLING DIAGRAM

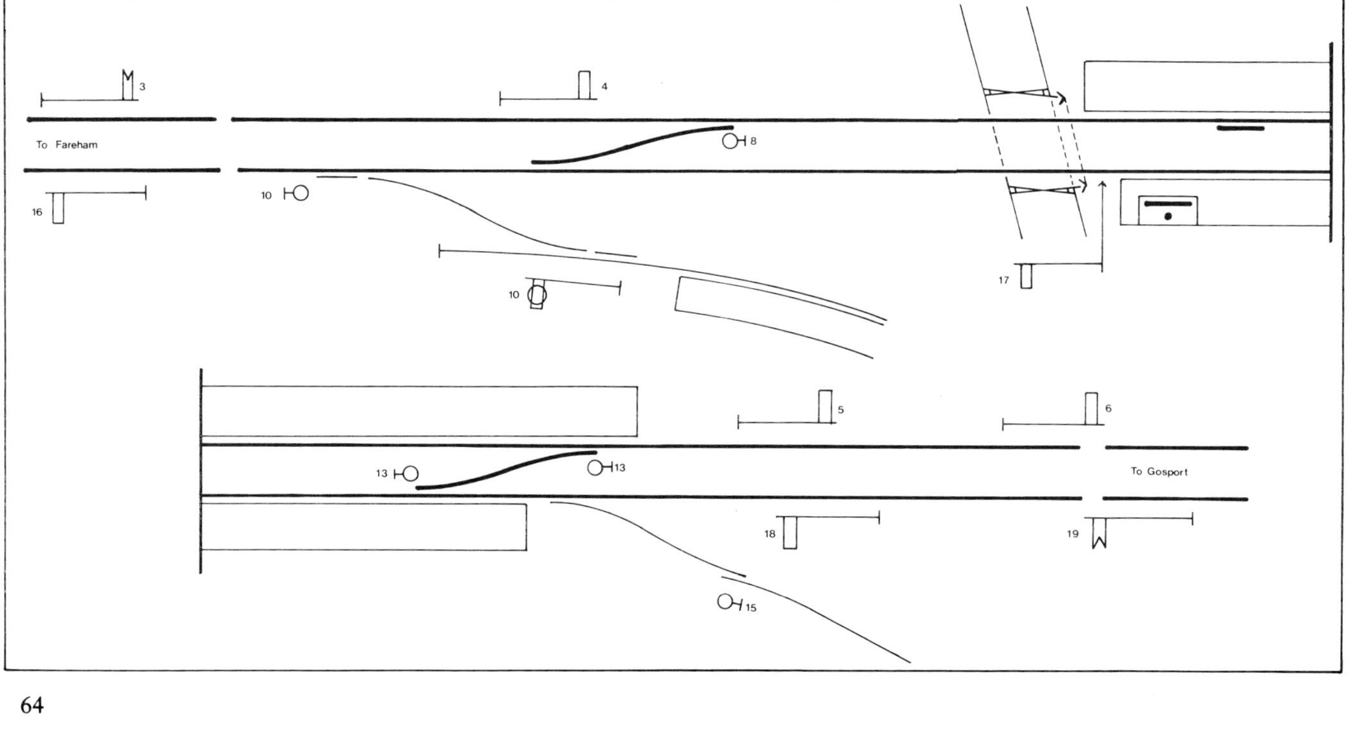

FORT BROCKHURST STATION BUILDINGS

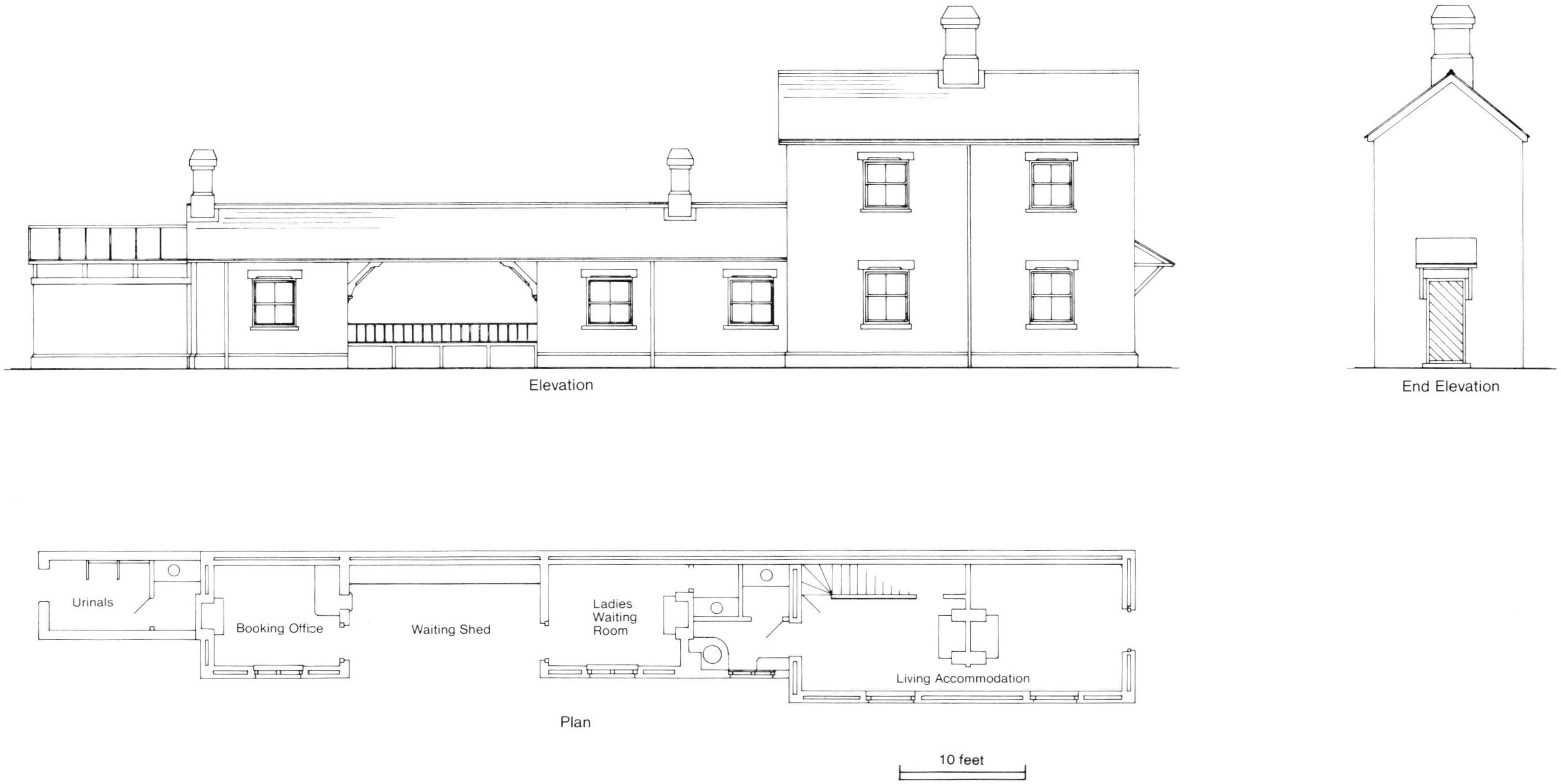

Fort Brockhurst Down Side Buildings

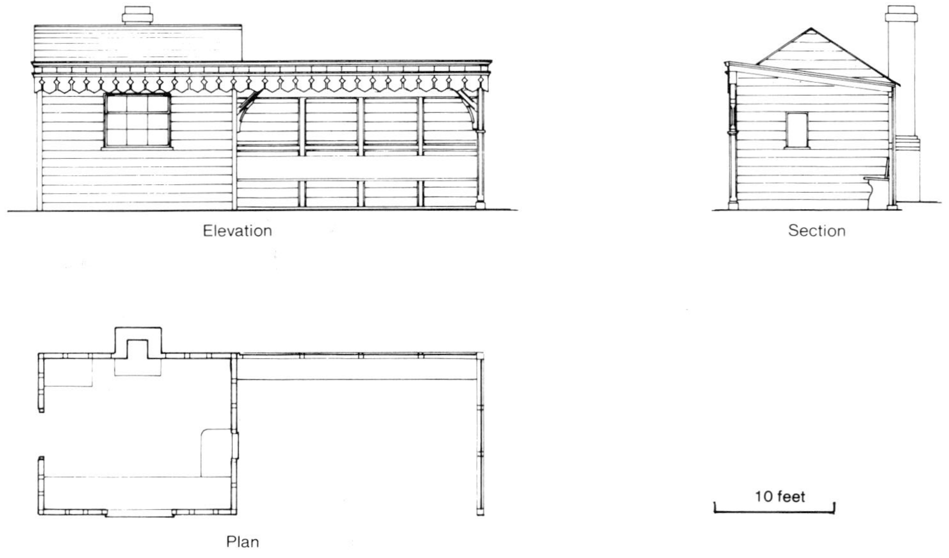

FORT BROCKHURST

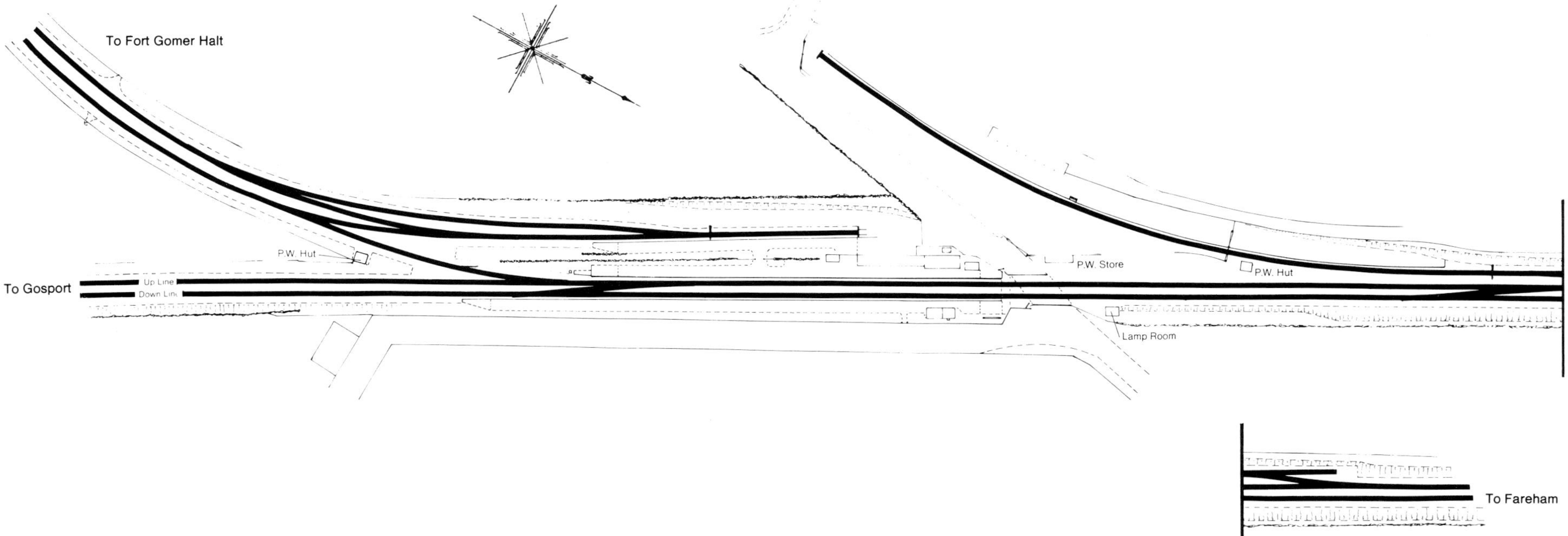

160 feet to 1 inch

Stationmaster 'Willie' Bew with his daughter and members of staff at Brockhurst around 1900. *Gosport Museum*

Proud railway staff pose alongside No. 21 *Scott* at the Lee platform. The pattern on the tank side is a result of polishing, probably with tallow. *Pamlin Prints, Croydon*

Another Lee line train at Fort Brockhurst with what would appear to be a mixed selection of stock. The gate between the main line and branch platforms is just discernable above the left-hand buffer. *R.C. Riley*

Staffing the Station

Despite the existence of a station house, early records indicate that the Gosport Superintendent exercised overall control of the new station at the time of opening. A Mr. Chandler was later appointed as stationmaster. He, however, was killed by a train entering the station when he fell from the platform edge. His replacement, a Mr. Lodder, was reputed to be a very keen gardener with the result that the topiary of the station was renowned far and wide. By 1895 control was under a Thomas Shepard, he moved on soon afterwards so that in May 1897 William Bew had taken charge from his previous post at Staines. Mr. Bew, or 'Willie' as he was known, was to stay at the station for some 26 years, earning a reputation as a proper English gentleman and respected by all around him. Upon his retirement in January 1925 a party was held in his honour at the nearby 'Artillery Arms', being presented with an armchair and pair of slippers, whilst his wife received a handbag and purse. Fortunately the guests at the presentation and concert in honour of Willie had not thought fit to remind him of a slight indiscretion some years previous when, on 25th August 1900, an engine was derailed at the station whilst crossing from one line to another. Mr. Bew was subsequently reprimanded for setting up single line working without making the proper arrangements and fortunately without causing further incident.

After a brief period, with no designated man in charge, a Mr. Hayward was appointed. His stay was destined to be brief, as in late 1929 he was transferred to Wool in Dorset with control passing once again to Gosport. This arrangement was to continue for the remainder of the life of the railway.

In addition to the stationmaster, there was for some years two signalmen, two porters and an office boy. Such a plethora of staff for a small station which was reduced in later years in the overall quest for economy of working.

Apart from the other incidents mentioned elsewhere, Brockhurst would seem to have had more than its fair share of difficulties. An example occurred at 10 am on 18th September 1917 when, during shunting, engine No. 745 was derailed at the points leading from the Lee platform to the run-round loop. The cause was put down to the locomotive wheels being worn. Almost exactly a year later, just after midday on 4th September 1918, the engine on the 8.30 am Eastleigh to Gosport goods was derailed at the west crossover points. This time it was stated to be due to an ambiguous hand signal having been given. Single line working was in force until nearly 5 pm, porter Batchelor and signalman Dyke being jointly responsible.

Other than the military traffic already spoken of, there was little freight handled. One exception to this involved the firm of C.H. House & Sons who, apart from being coal merchants, also sent market produce to London. The incoming coal traffic was received by railway wagon alongside the Lee line platform for many years. It is not known whether this firm had their own railway vehicles. Private passenger traffic too was minimal, especially when the competing street tramway came into being, although as was typical for the period, the booking office could offer facilities such as through booking to any station in the Kingdom as well as luggage in advance and other similar services.

After 1934 the station became a crossing place on the newly singled line between Fareham and Gosport, whilst later on with the closing of the Lee line to all traffic, the former wooden platform took on a new role as part of the west end sidings goods yard.

The impressive colonade and approach of Gosport station viewed before the fire damage during the war. At this time in its life it was nearly one hundred years old and seems to be in excellent condition. Stationmaster Ford is believed to be standing in front of the fence. *Lens of Sutton*

Gosport

Gosport Terminus owes its position to the requirements of the governor of the town in the nineteenth century. The railway was prohibited from entering the confines of the then fortified area and so had to content itself with a station destined to remain ¼ mile or so away from the preferred site nearer the commercial centre. The origins of the ornate station design have already been spoken of in a preceding chapter, although what may not be so widely known is that the railway company also had large offices in nearby Forton Road, these later used by Chaplins and still later by Pickfords as a furniture depository. The railway offices later moved to a new site adjacent to the station in Southampton. From an architectural viewpoint the renowned colonade is a fine example of early railway design and one of the few to survive into the mid-twentieth century.

Access for passengers was from the pleasantly sounding Spring Garden Lane, the station almost at the junction of Spring Garden Lane and Forton Road. Privacy of the site was maintained by a masonry wall flanked with railings, a pair of substantial gateposts fronting some equally massive wrought iron gates leading onto the station forecourt and cab hard-standing. The station buildings and colonade could hardly fail to impress the traveller, even if in later years it was victim to the ravages of time as well as showing the scars of enemy conflict.

For the intending passenger, entry onto the platform was through the booking hall. This was a cavanous chamber on one side of which were the pigeon hole windows of the booking office. Waiting rooms, including a ladies room, were provided within the main building together with the usual other offices and facilities. Incorporated in the end of the building was the station-master's house, the encumbent later provided with rented accommodation following the wartime bombings. There was also a garden for the stationmaster at the Fareham end of the station.

For almost all the life of the station, passenger trains arrived and departed from the one platform which was extended sometime around 1863. The opposite platform, the length of which remained that as originally built throughout its life, quickly found a use solely for goods and from about 1850 to around 1900 was designated as a cattle loading dock. Covering both platforms was a wooden and glass roof, this was supported on 1 foot square oak baulks until destroyed by fire in the 1940s. The overall roof was sufficient to cover four coaches, whilst the pervading smell that must have persisted around the immediate station area, consequent of having the cattle dock under the same cover as the passenger accommodation, is best left to the imagination! Indeed, this may well have been one of the reasons for a general change in goods handling in later years.

Around the turn of the century cattle handling was moved outside, the dock redesignated for goods handling and so rendering the existing goods shed superfluous. The old goods shed was then demolished around 1922 at a time when various changes were made to the yard layout.

Above: The Fareham end of the station showing clearly the old roof structure and taken in Southern Railway days despite the presence of an LSWR-marked wagon. The connection to the Clarence Yard ran from the centre line. *Pamlin Prints, Croydon*

Left: The inside of the station area showing the 1′ square oak supports. A good many parcels and assorted goods appear on the right. *Lens of Sutton*

Right: 'D1' No. 626 waits to leave with a Fareham service with unusually it appears four vehicles. The photograph was taken in the late 1930s. *Lens of Sutton*

GOSPORT 1850
160 feet to 1 inch
To Clarence Yard
To Brockhurst
Crane
Goods Shed
Ticket Platform
Engine Shed

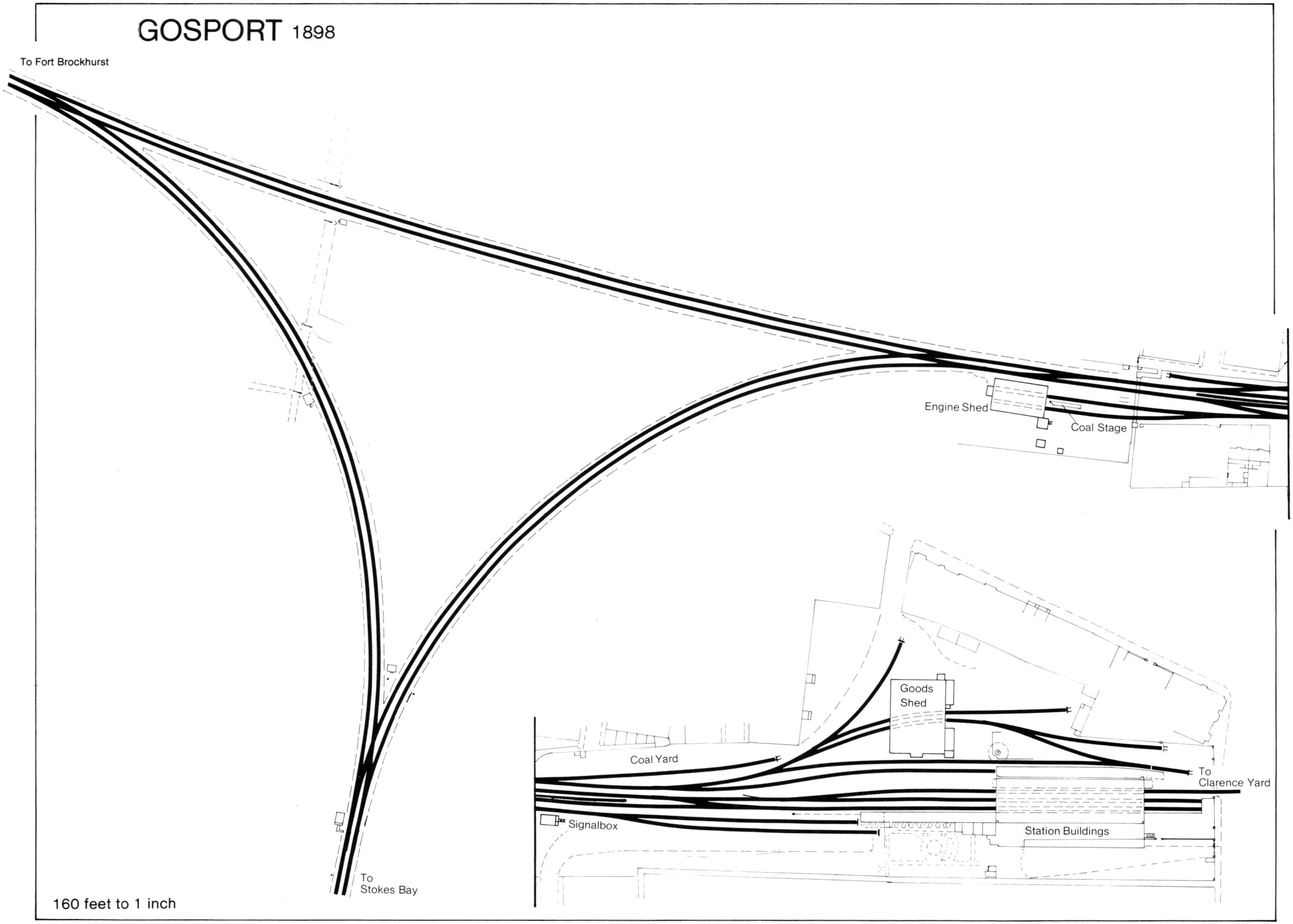
GOSPORT 1898
To Fort Brockhurst
Engine Shed
Coal Stage
Goods
Shed
Coal Yard
To
Clarence Yard
Signalbox
Station Buildings
To
Stokes Bay
160 feet to 1 inch

Above: The colonade, again taken in its heyday, with the photographer with his back towards the stationmasters garden.
F.E. Box collection, National Railway Museum

Right: Taken during the last years of the nineteenth century. It is interesting to compare this view with the previous two of the colonade, the only change appears to be the foliage. Just visible is one of the Chapman's wagons. *Gosport Museum*

Left: Another pre-war view of a Fareham train, taken probably in the late afternoon as the middle line is occupied with wagons no doubt ready for the evening goods departure.
Lens of Sutton

At either end of the goods platform there were various offices and warehouses, whilst direct unloading and loading onto road vehicles could be achieved via a succession of archways leading directly from the platform to a concrete loading ramp. It was local practice to chalk on eventual destinations of goods arriving at the station on the wall above the platform and so assist in sorting. Carts and road vehicles were able to reach the loading area through another equally large pair of iron gates fronting Spring Garden Lane.

Outside the station there were numerous sidings on both sides of the line, those on the 'up', or passenger side, primarily used for coal whilst general merchandise was handled within the main yard. The goods sidings were originally fed from a collection of wagon turntables, these being replaced between 1893 and 1898 by conventional pointwork. Examples of the track layouts at the station are shown on the accompanying plans. There was also a cartage roadway leading direct from the yard to Forton Road, whilst one of the yard sidings had over it a gantry crane for dealing with awkward shaped loads. A two-road brick-built engine shed just north of the station, on the 'up' side completed the layout.

Other than hand control of the points in the yard and sidings, operations were dictated from the brick signalbox at the Fareham end of the passenger platform. Previous to this early plans pre-dating the Stokes Bay line, show a 'switch box' between the 'up' main line and engine shed and situated just to the south of the latter. The later signalbox was a brick South Western design structure of the 'centre pillar' type and similar to numerous other 'boxes elsewhere on the system. It contained a 36-lever frame probably of Stevens' manufacture.

Looking reasonably busy, Gosport engine shed with the coaling stage on the extreme left. *M. Snellgrove collection*

Operation and Changes

Passenger services from Fareham would arrive at the platform and run almost to the buffer stops. Here the locomotive would uncouple and by means of an 'engine release crossover', run-round the train to couple up again ready for the return journey. Tender locomotives were able to be turned via the triangle, whilst it was often the practice to use the locomotive for shunting purposes in the lay-over before departure time. In comparison, goods services were dealt with by the locomotive

'A12' 0-4-2 No. 604 on what is probably an Alton service in August 1931. Note the private owner wagons on the adjacent line.
L&GRP courtesy David & Charles

running round so as to leave the brake van in the middle road of the station before pushing the wagons to their respective positions in the yard. On occasions, when no passenger trains were due, the passenger platform was also utilised for the temporary stabling of freight stock.

In addition, horses were for many years used to assist with the shunting, Chaplin & Co., the well-known South Western carriers, having a warehouse and stables adjacent to the yard from which the animals were supplied as required. The firms local office was at No. 55 High Street. Parcels and the general collection of goods in the vicinity of the town were also under the control of the delivery agent, for much of the time a local man using a horse and cart. In June 1859 for example the agency was reported as having passed from Mr. Hysop to Mr. Nance, consequent upon the death of the former.

Old plans also show a separate small stable on the railway property itself, this was where horses were once kept and then used to move wagons around within the old goods shed. Bill Westbrook remembers being charged with their keep and use. Next to this was the old harness room, later used for timber

Above: One of the occasional Gosport-Netley services waiting to depart from Gosport behind '700' class No. 350. This service was primarily for the use of schoolchildren.
F.E. Box collection, National Railway Museum

Left: Right up until the final passenger service in 1953 the stock comprised mainly of compartment stock. In this view the stock is waiting in the platform whilst the locomotive appears to be shunting in the station approaches.
H.C. Casserley

Above: Gosport station seen from Spring Garden Lane in September 1952. The station now sports its new roof although by this time the whole scene looks rather sorrowful having been left to deteriorate.
H.C. Casserley

Left: The station viewed from the Fareham end showing the corrugated building on the left which was used as a grain store. The roof covering of the goods side only can be clearly seen. *Lens of Sutton*

storage, whilst the former stables were converted to offices for the goods department staff.

As would be expected with any railway locality, various changes took place over the years in line with traffic and operational requirements, both local and national. It would be both unnecessary and monotonous to attempt to list all so a random selection has been made of some of the more interesting items.

Of these the first must be in 1863 when the curve to the Stokes Bay line came into use and altered soon afterwards to become a triangle with the opening of the opposite connection. Little is known of the workings of the period, although an 1865 entry in the minute book does confirm that signals were at the time provided at both Gosport and Stokes Bay Junction. It was recommended that the signals at both localities be lit by gas. Then in 1877 changes were made to the living accommodation for the station agent. Two extra bedrooms were provided as well as changes to the kitchen. The cost of these was put at £300.

It would appear from contemporary records that the railway company directors made periodic inspections of various installations, one of these to Gosport around September 1883 as a result of which £85 was authorised to be spent on partitioning a portion of the existing booking office so it could be used as a general waiting room. Around the same time fire extinguishers were supplied for the first time along with three supply hydrants, the cost of this put at £97. A contemporary account of both changes reporting thus;

> '..... the conversion is now nearly completed and already the clerical staff have entered into possession of the smaller, though sufficiently commodious, booking office. This is situated near to the entrance gates and beyond are the Ladies Waiting Room (which formally existed but to which access is now gained from the platform) and the new General Waiting Room. Another improvement is the provision of a water main which runs around the building, this being connected with three hydrants, ample means are thus provided for coping with any disaster which might arise in the matter of fire .....'

One may perhaps wonder if there had already been a fire at the station.

GOSPORT New Goods Offices

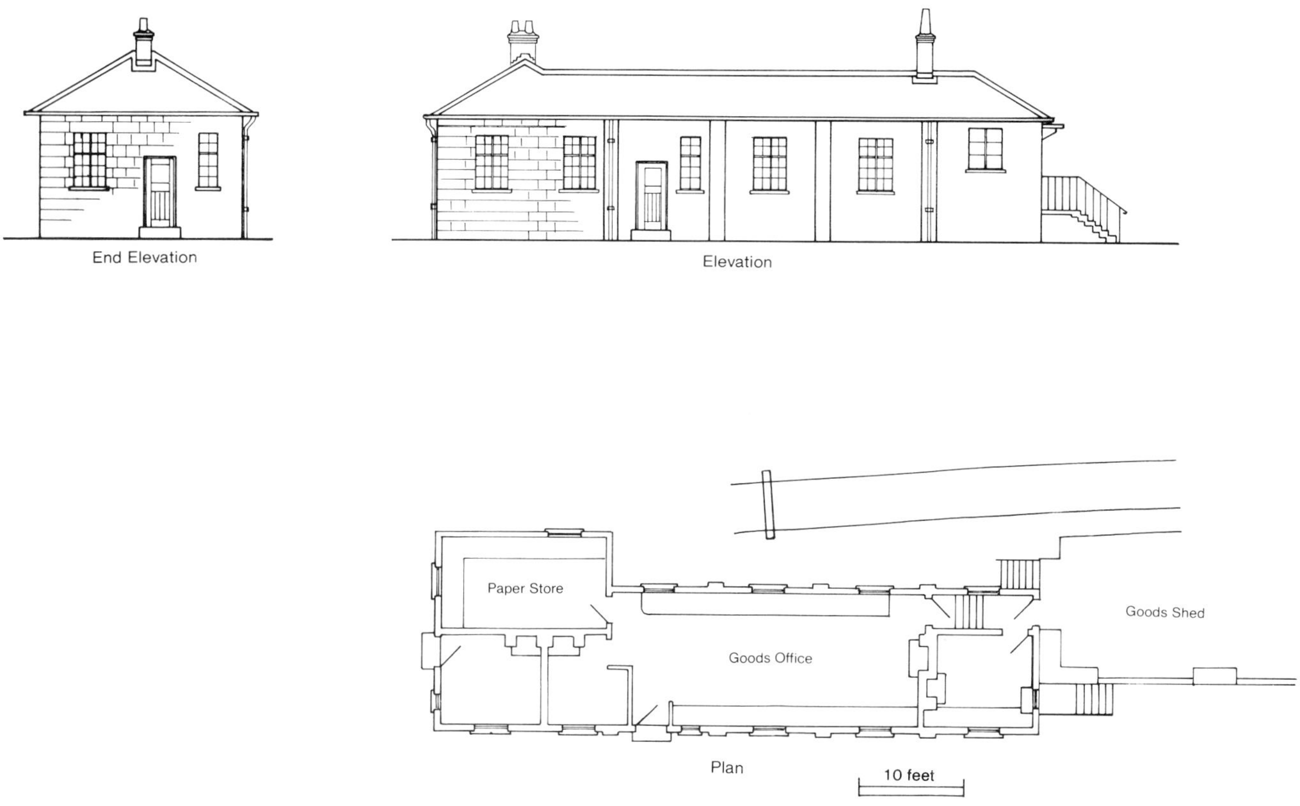

Unusually busy times at the station with much in the way of goods traffic visible. In fact the day in 1953 was the occasion of a special train to Gosport which had previously visited Bishop's Waltham. *P.J. Kelley*

A large number of references also relate to the goods side. The various improvements in available facilities a good indication of what must have been a considerable increase in traffic handled. The first of these was authorised on 29th January 1890 with an extension to the goods office costing £68, although it is not clear whether this refers to the goods platform or outside goods shed. The next year £214 was spent on a new 20 ton weightbridge, destined itself be replaced in late 1943 with a secondhand larger 30 ton capacity item. The cost on the latter occasion having risen to £924.

Separate cattle pens were apparently first provided sometime after 1850, these were paved from around 1899. Additionally in 1921 a concrete hardstanding with water supply and drains was installed at the end of the goods platform so that cattle trucks could be hosed down after having disgorged their occupants.

As mentioned above the wagon turntables were removed towards the end of the last century, this just one of numerous alterations to the track layout affecting the goods side over something like a 25 year period. In addition, in 1895, additional goods accommodation was provided at an estimated cost of £2,195 for new sidings. Tantilisingly though the sidings concerned are not detailed. A further £178 was spent again on additional sinding(s) in 1902.

Later in 1915, and reported to be due to an increase in goods traffic, a made-up roadway was provided alongside one of the yard sidings. It was anticipated the hard standing would afford easier unloading facilities.

The largest increase though came in 1922, with the directors authorising no less than £11,250 to be spent on changes together with a further £825 on a yard crane. Unfortunately no details are recorded although it may be reasonable to assume the work included removal of the yard goods shed and consequent realignment of the various sidings. Finally with the new ownership of the Southern Railway but six months old, £153 was spent on enlarging the concrete standing around the exterior cattle pens and loading dock.

On the passenger side changes too are listed, although the seeming lack of importance reflected on this aspect of traffic is perhaps shown by the few items recorded. Indeed only three references appear over a number of years none of which were likely to have much bearing upon the operation of the terminus. The first was in November 1897 when footwarmer apparatus was provided for passengers at an estimated cost of £118. Then in 1915 a wash basin deemed surplus to requirements was removed from the ladies wash room and finally on 28th March 1916 authority was given for the ticket platform just outside the station to be removed, (this believed to have been altered in position over the years).

A consignment of chains comprising some 4,000 10 inch links for the Floating Bridge in 1932. *Portsmouth Evening News*

Traffic Handled

The staple trade of Gosport station was for many years coal. By 1860 for instance there were no less than 10 merchants operating in the area. Many of these had wharfs within the station yard. Some of the firms recalled are the Gosport and Alverstoke Fuel Association, Co-Op, Messrs. Smith, Renwick-Wilton, Stevens & Clark, Clement Bros., Bryer Ash and A.C. Barnes, several of these thought to have at one time operated their own private owner wagons, whilst others rented vehicles belonging to other local merchants, not always from the same town. Barnes was probably the last man to use a horse for coal deliveries around the town, the practice continuing until the horse died in the late 1940s.

There was also much freight traffic to and from the Isle of Wight. This was sorted outside the main station before being dealt with by Chaplins who had there own lighters and barges for onward transmission. As already mentioned, there was also considerable naval and military traffic, personnel travelling in both directions as well as incoming stores and cattle 'on the hoof'. Naval and military hardware was also dealt with resulting in several exceptional loads. Some of these were recorded within the 'Southern Railway Magazine', examples of which are shown on the accompanying pages.

Being one of the earliest railway stations, Gosport had been witness to a series of radical changes to the way in which passengers were dealt with. One of the most novel ideas was when huge bells were hung between the lintels atop the chimneys and rung for five minutes before the train departed. One can only imagine the resultant din. Many years later the legacy of this was still visible in the space between the chimneys on the station.

A record load of timber – 115 feet long consigned from Nine Elms to Gosport in 1926.
Southern Railway Magazine

A 50 foot long launch loaded ready for its journey from Gosport Yard to Newcastle.
Southern Railway Magazine

Among The Staff

Right from the earliest times, Gosport had a superintendent in charge. This man was responsible not only for the station and its traffic but also for the employment or otherwise of staff in the vicinity. The first of these is thought to have been a Mr. Watkins, who later moved to take control of Southampton. By 1852 Richard Stevens was in control and by 1875 John Madigan, the title by this time having changed to that of Station Agent. A further change had occurred by 1886 when Joseph Dyson is mentioned, the position of Station Agent Gosport then including control of the neighbouring stopping places of Gosport Road, Stokes Bay and Brockhurst.

Mr. Dyson left around 1895, after which records become somewhat clearer. The new agent was Joseph Garnham who remained in charge until 31st December 1911. Gosport by this time was graded as a 'Class 2' station and Alfred Capel was appointed to the post as a replacement at a salary of £145 pa. Capel was replaced in 1920 by Alf. E. Geary, records showing that inflation during the war years had taken its toll, for his salary was no less than £300 pa, minus £40 for rent of the station house. Some accommodation was also available for other railway staff in a pair of cottages to the rear of the signalbox, the senior shed driver occupying one of these for many years.

Mr. Geary left in March 1926 to take up a similar position at Wickham, the move indicating the lack of importance even then being placed upon the station's grading. Records are then confused slightly for there is a brief mention of a Mr. A.V. Jeffries before perhaps the best known of all the stationmasters arrived in January 1927, Horace C. Ford, recalled as always sporting a carnation in his button hole.

Mr. Ford was destined to remain as stationmaster for less than five years, but during that time he made his presence felt in many ways, not least his abrasive manners and persuasive character. Perhaps the most lasting result of this latter trait was the opening of the Gosport Station Club on 15th March 1930 in what had been the old parcels office. This was basically a private licensed club for the benefit of the railway staff and run by a full time steward. Its formation was reputed to have come about following an argument between Mr. Ford and a local licencee after which Mr. Ford refused to enter the same public house again, instead proclaiming he would open his own, which is of course what he did. Both food and drink could be obtained, whilst for relaxation, card and billiard tables were installed. One may only wonder as to the autonomy of the stationmaster in obtaining official sanction for the venture.

Mr. Ford left in August 1931, on promotion to station-master at Barnstaple, one of his final acts was to invite a senior porter, Jim Prior, to the nearby Railway Hotel for a drink. Prior, not only had to pay for the round but lost the change which Ford pocketed. Another incident recalled is the collection that went round when Mr. Ford was leaving. This was started by Ford himself with a donation of 10/-! His replacement was Mr. M. Searle who remained in charge until November 1941.

The final two encumbents under Southern Railway auspices were both men who had previously worked as clerks at Gosport; Mr. O. Ponting who had moved from Gosport to Woolston and Sholing on promotion before returning and then from November 1945 Mr. E.J. Pond. Interestingly the 'Southern Railway Magazine' recorded Mr. Pond as taking charge at Gosport, Fort Brockhurst and Lee, even though the latter place had by this time been closed for almost a decade. After British Railways had come upon the scene a man from Haslemere was placed in charge, he is recalled as being unpopular due to cutting out all staff overtime!

In addition there were various clerks, foremen, porters, shunters, checkers, office lads, guards, carriage cleaners and signalmen working at the station as well as, of course, drivers and firemen. The number of staff varied over the years to suit changing traffic needs. The carriage cleaner was a night shift worker whose task it was to sweep out the carriages which were stabled in the platform overnight. In addition, the vehicles were meant to be washed externally on both the platform and track sides. Another of his allocated tasks was to check on security of the station during the quiet hours when no trains were running. Often though he would rush to complete his cleaning so as to be able to spend the rest of the time asleep in one of the coaches.

As was typical of the period there were several staff who completed fifty or more years service, one of these was Mr. A. Williams who in April 1929 retired from the position of crossing keeper after fifty-two years service. Another, the Chief Goods Clerk, Mr. C.V. Botton, finished at the end of 1927 after fifty-three years service and yet another goods clerk F.C. Cox, who retired in October 1928 again with over fifty years service, nineteen of which had been spent at Gosport.

From about 1935 onwards women were taken on to the staff, one of the first of these was in March of that year when Mrs. M. Bailey was appointed crossing keeper at Gosport, unfortunately the actual crossing referred to is unrecorded. A little later, in the 1940s, others were taken on to replace men serving

The magnificent colonade showing some of the staff and a delightful horse carriage waiting for business. *Gosport Museum*

Inside the colonade c. 1890. *Gosport Museum*

with the armed forces, this time working in a variety of posts including that of clerks.

Other than the use of the clubroom, there were few legitimate benefits the staff could obtain, although tips from passengers was a welcome bonus. One exception, however, was the practice of being able to purchase old sleepers for 6d each as firewood, although the heavy creosote content was likely to tar the domestic chimneys in no time at all.

The Locomotive Department

Just north of the station and again on the 'up' or Fareham side of the line, was a two-road engine shed. This is believed to date from the time of opening of the railway in 1841. The shed itself was of brick under a tile roof. Interior smoke chutes vented to a central wooden slatted belvedere. The shed roads were both 'dead ends' and scaled up from plans, provided cover over a distance of some 65 feet. At the station end of the shed there was also a water tank and crane whilst at one time a wagon turntable led off at right angles to a short siding likely to have been used for coal wagons. A pump house near the end of the stub siding fed the water tank, adjacent to which was a small office.

By 1893 the 'stub' siding had disappeared and a coaling stage on the site of the former switch box is shown. The coal stage was later moved to a more logical position outside the shed and between the two lines. Later still, around 1917, a short siding appears running off the shed and near to the boundary fence, a 'V' of land resulting, again used for coaling from a wooden stage. It was practice for locomotives to draw up alongside this stagework after which they would be hand coaled either from the stage or direct from the wagon, in either case a considerable amount of manhandling of the coal was needed. Plans also exist around the same time showing proposals for this siding to be extended to connect into the 'down' Stokes Bay line and able then to hold twenty-two wagons. A 30 foot access roadway would also have been provided alongside. Such proposals were not, however, effected.

'0395' class 0-4-2 No. 0442 shunting in the goods platform on 8th September 1928. *Pamlin Prints, Croydon*

Within the shed the inside walls were whitewashed at intervals by one of the cleaners, whilst examination pits were provided along with a work bench running down one side. Only running maintenance could be carried out, major repairs being undertaken at the parent depot of Fratton, whilst a fitter or boilersmith would be sent over as the occasion demanded.

The shed had a number of resident crews, six for example in the 1920s although diminishing traffic later reduced this

Another 0-4-2 this time No. 644 on a Fareham-Gosport goods. Just outside of the shed can be seen a bin of pebbles which were used by the crews to prevent the clinkering of the firebars. *Lens of Sutton*

number by half. One of the drivers was appointed as 'in charge' and responsible for ordering oil and stores etc. as well as general welfare, he received an additional 10/- weekly as compensation. In addition there were a number of cleaners, three in 1920 but only one by 1950. The cleaners also acted as fire raisers and general shed labourers charged with coaling the locomotives as well as cleaning them and site work. Relief staff were provided from Fratton as required although Gosport was not always a popular location with certain of the relief crews.

The elegant 'T9s' and 'M7s' were probably the most regular performers at Gosport for many years. Here Nine Elms-built No. 30336 shunts the yard in 1952.
H.C. Casserley

Accidents

When compared with Brockhurst, Gosport has little recorded in the way of incidents in the early years. Indeed, in the ten year period from 1910 to 1920, only two such situations are mentioned. The first, on 16th June 1913, was when tank engine No. 3 was derailed at the cross over points at the Fareham end of the station whilst in the course of running round its train prior to proceeding to Stokes Bay. Three trains were disrupted as a result and ran direct to Stokes Bay rather than reversing in Gosport station. Ordinary Fareham – Gosport services were terminated at the ticket platform outside the station.

Then at 3.15 pm on 20th May 1919, two wagons were derailed on the main line whilst in the course of being made up as part of the 3.40 pm Gosport to Eastleigh goods. The cause was stated to be insecure trusses on a hay wagon with two of the yard staff being at fault. Two passenger trains in either direction had to be cancelled.

GOSPORT SIGNALLING DIAGRAM

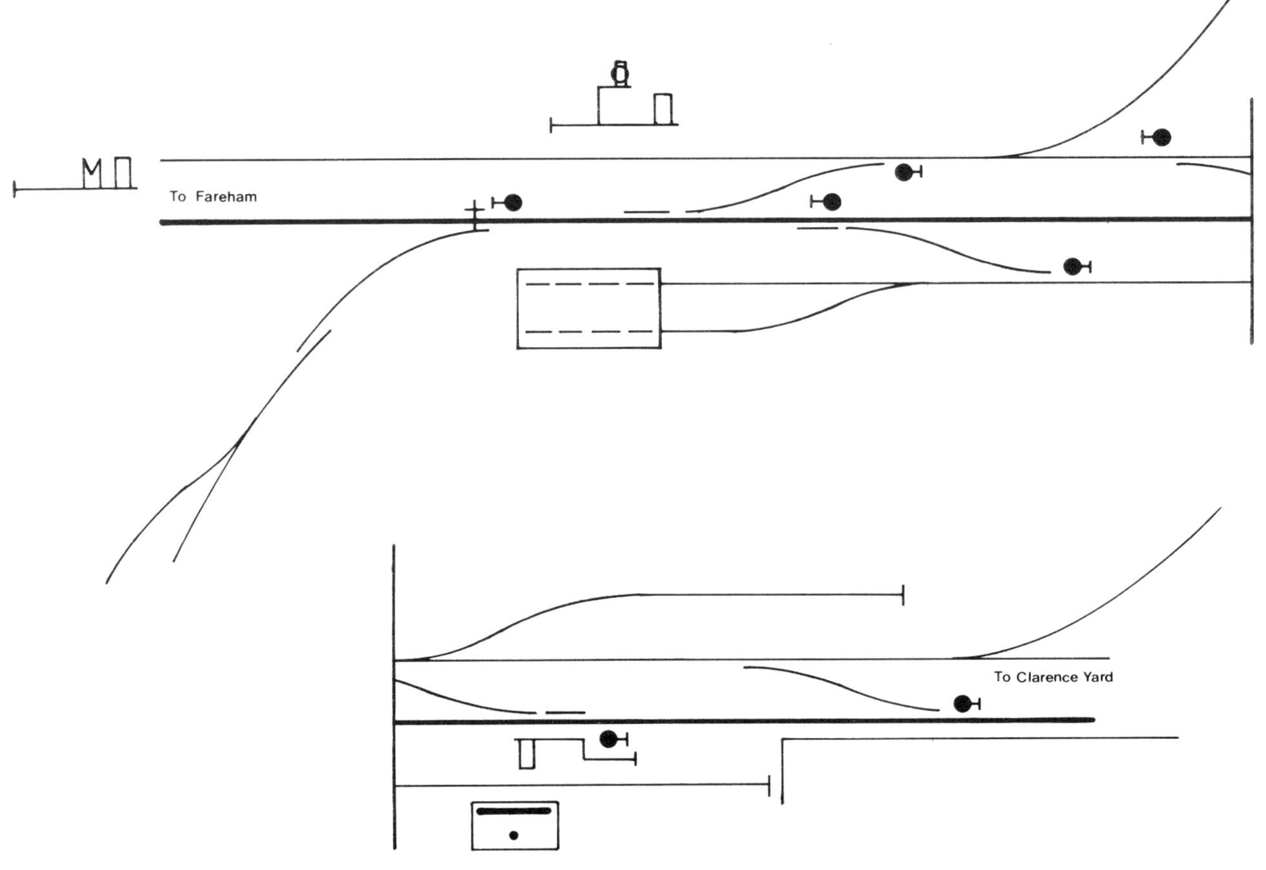

Gosport Road

Of all the stations in the area, Gosport Road is one of the more obscure. It has a particular fascination and one which no doubt stems from the lack of available information.

As recounted in chapter four, the station was opened on 1st June 1865, the same time as the east curve allowing trains direct access from Fareham to Stokes Bay and so avoiding the need for a reversal at Gosport station. It was then known as 'Stoke Road' station, this name derived from a nearby thoroughfare. Its purpose was to allow passengers destined for Gosport itself to board Stokes Bay trains and so alight still within relatively easy access of the town. Cabs were available to meet the services.

How long the original name was carried is unclear, for the following year 'Gosport Road and Alverstoke' appears, whilst the twin arch bridge carrying Bury Road at right angles over the railway, whilst at the south end existed the trellis bridge mentioned above.

There was no yard or siding of any type, although a signal-box was later provided, probably during the mid 1880s. Consequently, only passenger traffic could be handled, this limitation being mentioned in the Railway Clearing House handbook of stations.

Despite being opened at a time when the Stokes Bay line was still a nominally independent concern, it is likely that the South Western were instrumental in arranging the facilities as well as the staffing. Records indicate that throughout its life control was exercised from Gosport itself, a porter being

A superb rare view of Gosport Road station c. 1868 looking south towards Stokes Bay. The photograph was taken following a number of complaints being received by the railway company concerning facilities and services. *British Rail*

from November 1866 the title 'Gosport Road' alone was used. In the past this latter date has also been confused with the actual time of opening.

The station itself consisted of 'up' and 'down' platforms, these being low level and originally surmounted by a brick waiting shelter on either platform which was itself open on one side. In addition, the 'up' line, or Fareham bound platform, had a wooden ticket office. Public access to the platforms was via a set of steps on the 'down' side and a path on the 'up' side, the two being connected by a wooden trellis bridge which also served as a public right-of-way. At the Fareham end of the station there was outstationed who would act as ticket collector, cleaner, porter and clerk. As a matter of interest a 'Station Agents' office is shown on the plan of 1896.

It is likely that traffic receipts were small as local residents tended to only use the station in the 'down' direction when a trip to the Isle of Wight via the Stokes Bay ferry was planned, although bearing in mind the short distance involved, there were many who walked.

But even if receipts and therefore the facilities, were small there were several complaints about the lack of trains between Gosport Road and Fareham. One, in late 1868, when a

GOSPORT ROAD 1900

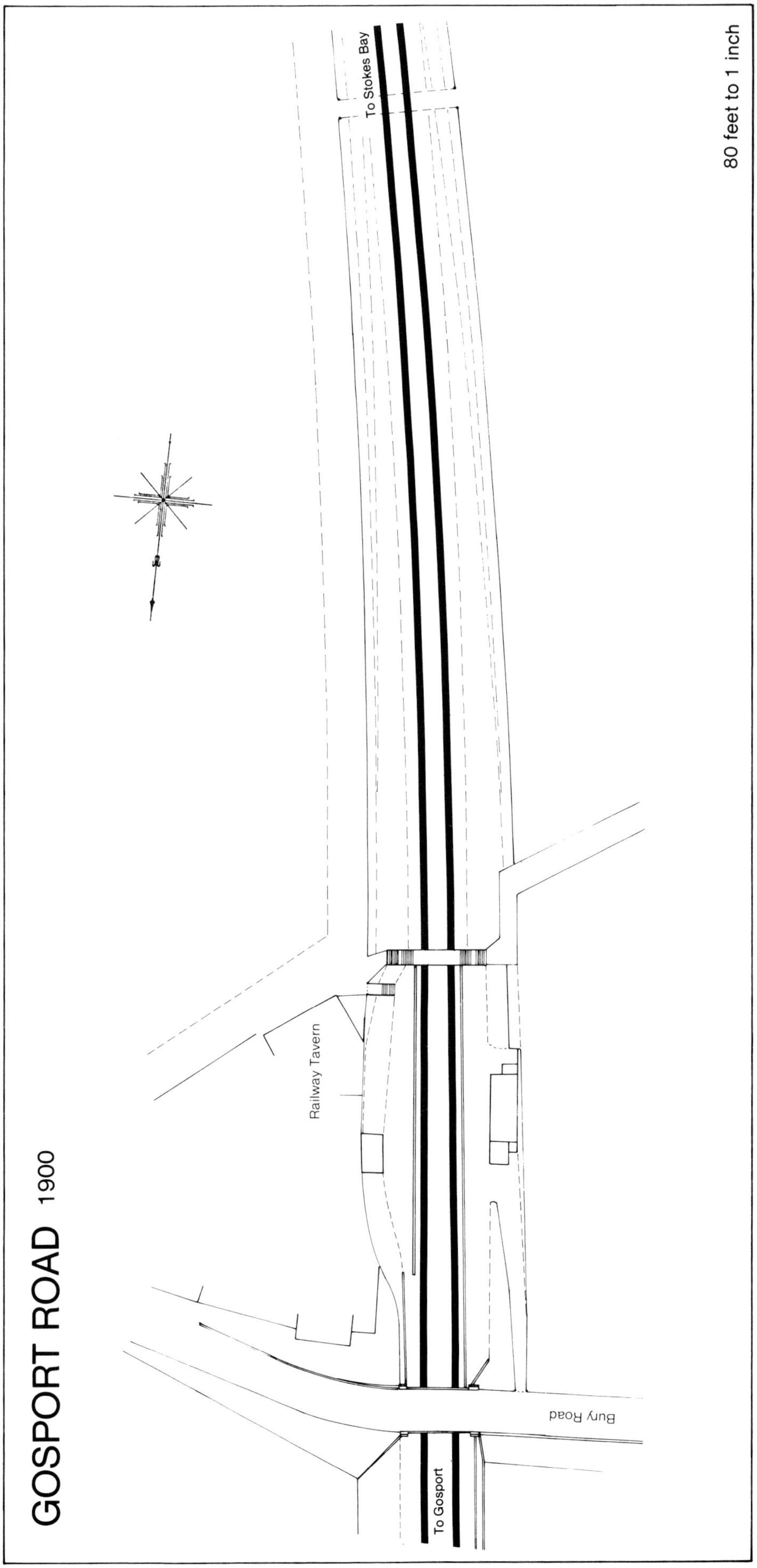

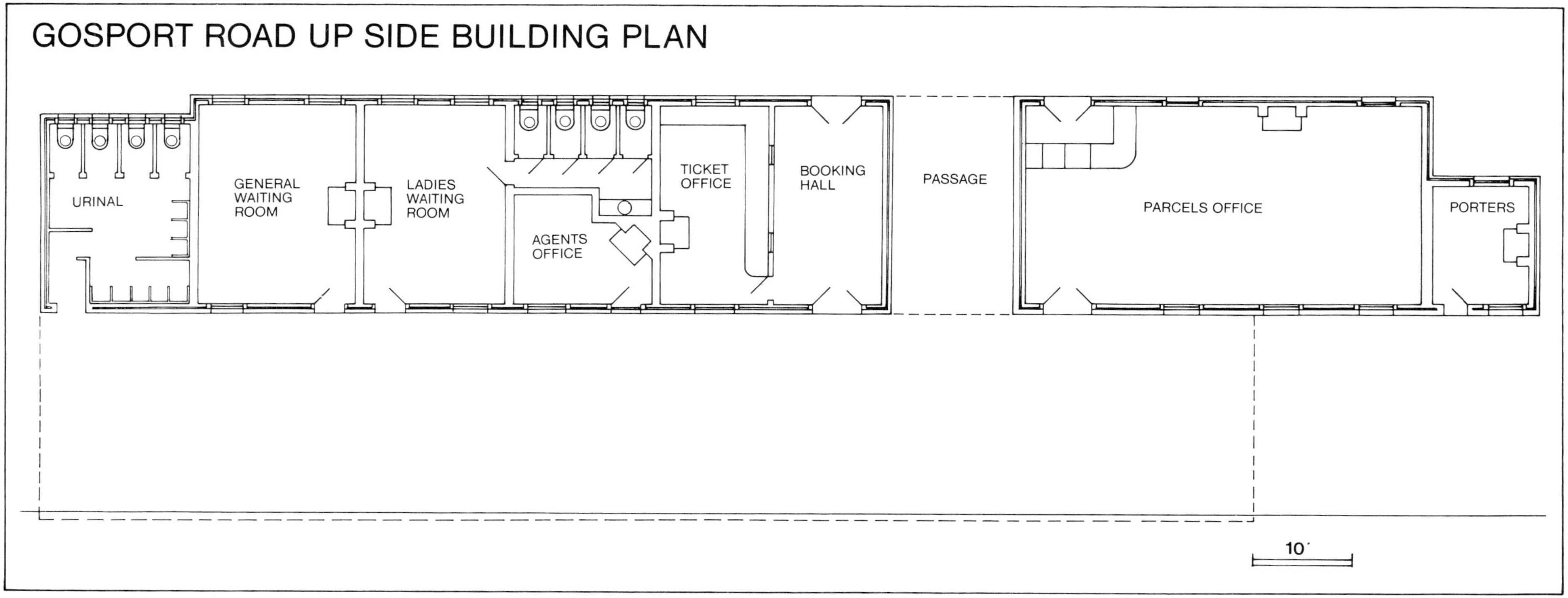

Looking north through the station towards Gosport and the triangle. The twin arches of Bury Road bridge can be clearly seen beyond the footbridge. *Lens of Sutton*

The station approach and cab stand with the wooden gates which allowed access to the 'down' side platform. *Gosport Museum*

An unknown lady on the footbridge which was a right-of-way. *Gosport Museum*

'memorial' was sent by the residents of Gosport Road to the Board of Trade complaining not only of the train service but also about the lack of station facilities. One may ponder as to who exactly these residents were, for at the time there was little if any residential development within the area. Despite this the Board of Trade were suitably moved to inspect the station, although correspondence from the Inspectorate doubts if they can do much to improve the service.

Accordingly, a re-inspection of the site was made, and as a consequence provided valuable information for this history like the size of the 'up' side shelter at 19′ x 13″, the 'down' side shelter at 17′6″ x 8′6″ and the existence of a booking office and wooden hut. The approach to the 'up' platform was stated to be via a path some 30 feet long. The report commented that the South Western should consider building a proper waiting room on the 'up' side along with a water closet.

No reference has been found to indicate that the South Western may have complied with these suggestions and unfortunately no photographic evidence assists. Changes were made later as a note appears in the minute book for 18th February 1885. It was reported a request had been received for increased passenger accommodation as well as additional trains to and from the station. Unfortunately the name of the originators are unrecorded, could it perhaps have been the local residents again? The railway company postponed a decision, although the request was subsequently approved on 1st April 1885.

Later matters began to move towards a modification of resources. A plan submitted by the General Manager on 9th December 1885 outlined the enlargement of the booking office together with a proposed carriage drive to the station. The cost of these improvements was estimated at £148 and £80 respectively. The need for an increase in booking office accommodation having first been mentioned in October of the preceding year. It is not clear if these works were actually completed, although on 31st March 1886 reference is made to expenditure of £26 on improvements to a waiting shed and although not specified, it is likely this referred to the 'down' side.

It may be reasonable to assume that sometime soon after this, major rebuilding work was carried out, both platforms now

For many years a 'hump' in Bury Road indicated the position of the railway line. The only recognisable feature today is the White Hart public house, once alongside the station. *Gosport Museum*

being supported on brick and of standard height whilst on the 'up' side a single storey brick building was provided. Details of which are shown on the accompanying plan. The 'down' side retained a solitary waiting shelter and from plans was somewhat smaller than had originally existed. It is not thought full platform canopies were fitted, although certainly by the end of 1896 they had been added at a cost of £140. Around the same time the original access footbridge was altered. This was now a single steel girder, secondhand from near Romsey and having a span of just under 50 feet. It was supported on brick piers and positioned slightly north of the original bridge.

The signalbox referred to above was on the platform at the north end of the 'up' platform. It was not a 'block post' and instead contained a frame of six levers, primarily as a ground frame for the boxes on either side at a time when severe limitations existed on the maximum distance a signal could be worked. In 1901, in conjunction with the reconstruction of the viaducts on the Stokes Bay line, the box was upgraded to a block post and reported to be temporarily moved to the south of the 'up' platform. At the same time three additional levers were fitted to the frame and a tablet instrument installed. A crossover was also provided south of the station to allow trains access onto what was now a temporary single line in connection with the engineering works. It was stated that when the works were completed the box would be down graded to its original non-block status and the double line restored, at the same time resiting the structure as before. Practicably, however, it may well have been that the temporary position was occupied by a temporary signalbox, such practice a regular feature of Southern working in later years.

Finally, in July 1922, at a time when the line and station had been closed for several years, approval was given for the 'up' side buildings to be converted into living accommodation for guard Belbin who was unable to find a house elsewhere. Cost of the conversion was estimated at £437 with the rent set at 12/- weekly.

Signalman Ted Gates, for many years at Stokes Bay Junction, surveys the scene at Gosport Road in about 1930. A sleeper in the distance on the line indicates the extent of the line then used. *Lens of Sutton*

GOSPORT ROAD SIGNALLING DIAGRAM

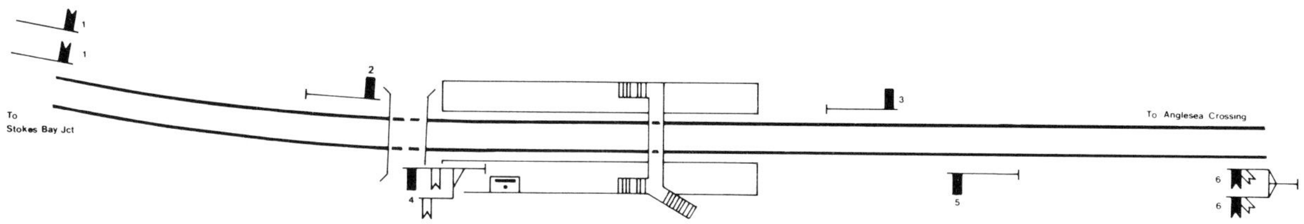

Stokes Bay

Stokes Bay pier from the shore on a rather blustery day. This type of weather no doubt made it quite hazardous to the ferry operators. *Sean Bolan*

The terminus at Stokes Bay was situated on a pier over the sea and having only limited local access to what was intended primarily to be a station dealing mainly with travellers to and from the Isle of Wight.

The hopes of the promotors are illustrated by the expanse of facilities provided from the outset, not least of which was the double track leading to the terminus even if only a single line with double platform face was provided. The following extract from a contemporary periodical describes graphically the early situation;

> 'The pier which it is reckoned will be accessible for steamers at all seasons and in any weather, is about 600 feet in length. On it is a commodious shed, capable of receiving a train of five railway carriages, so that passengers can step out of the train under shelter and close to the waters' edge. There are also commodious waiting rooms for ladies and gentlemen and the requisite offices, while the pier head is admirably adapted for a promonade and has a superficial area of nearly 4,000 square feet available for that purpose.'

The offices and rooms spoken of above were located at the end of the line of rails, adjacent to which a ramp led down either side to the pier itself. This had three sides, each with a landing stage, although it is unlikely all three were required to be in use at the one time. The various rooms, as well as the pier itself, were illuminated by gas. The reference to a 600 foot length pier is accepted to refer to the combined moorings available, whilst there is no record of it being used as a promonade as had been suggested.

The second line of rails from Gosport Road had a crossover to afford access to the platform, after which it continued forward a short distance in the form of a siding and with a crane of 1½ ton capacity alongside. The station is mentioned as being able to handle goods and according to early editions of the L & SWR timetables, having an express goods service to numerous important centres. In practice though the facility was available only if required, although passengers' luggage was probably the only regular freight handled. Certainly there is no record of separate goods train being run, the odd wagon that did arrive was attached to the ordinary passenger service or worked as a special transfer working from Gosport.

As with Gosport Road, control of Stokes Bay was exercised from Gosport, this arrangement continuing throughout its life. Staff consisted of a ticket collector, crane man and night watchman for the pier, these were supplemented as required by additional staff from Gosport. Indeed the line was popular with staff for two reasons, firstly the opportunities for fishing and also the tips they could acrue. It was not unknown for a man to be able to live off his tips rather than his wages during the busiest weeks of the year. In May 1899 it was reported that the railway staff at Stokes Bay were unable to find house accommodation within a reasonable distance of the station and so it was recommended that a block of five cottages be built for a estimated cost of £1,235. Two of these would be for members of the traffic department with the others for engineering department staff. No mention is made of the arrangements that had existed between 1863 and 1899.

The pier and station structure did not lend itself to much in the way of alterations and consequently few changes are recorded. One alteration though was the provision of a ticket platform just north of the pier in August 1863. Sanction for this was given at the same time as the similar facility at Gosport was lengthened. Several years later, in June 1880, reference is made to the provision of a refreshment room on the pier, but no decision is recorded. The refreshment room proposal was brought up again the following year, together with a request for the platform to be lengthened. Records show plans of both were prepared but not acted upon. This indicates that traffic was increasing but not in sufficient quantities to justify the expansion.

The single siding alongside the platform and laid with track spiked directly onto the sleepers. *Sean Bolan*

The biggest change during the life of the pier as a railway connection undoubtably came in 1895. An examination of the supports revealed that considerable work was necessary to the iron superstructure, primarily in the vicinity of the platforms and track. The directors had little choice but to approve the repairs if they wished to retain the use of the line and so approval was given for the funding of some £6,610 to cover costs. Before the work could commence suitable arrangements had to be made for dealing with trains other than at the existing platform. It was decided to construct a temporary platform 420 feet long just north of the pier and connected by a temporary viaduct. The track and signalling layout was also altered, whilst alongside the new platform was a siding able to be used as a stacking ground for the replacement girders as well as having an unloading crane. The connection to the original platform and siding were then severed and as no buildings are shown on the temporary structure it may be assumed that the original offices remained in use.

The Board of Trade were informed of the new arrangements on 12th March 1896, the temporary work being completed and shown as awaiting inspection three weeks later. How long the pier remained in this form is uncertain.

As intimated, control of the railway at the pier was exercised from a small signalbox located on the 'down' side of the line just outside the station. Few details of it have apparently survived, the only signalling plan found showing the modifications applicable to the temporary rebuilding, although it is likely to have been a wooden structure. The telegraph system of communications were installed on the Stokes Bay line from early 1864.

The railway company were also under certain obligations to secure passenger safety at Stokes Bay and had to provide a boat for life saving purposes. This was reported in 1911 as being unseaworthy and past repair and a replacement was provided by

This hand operated winch was positioned at the end of the siding and supported in a somewhat precarious manner. *Sean Bolan*

STOKES BAY PIER

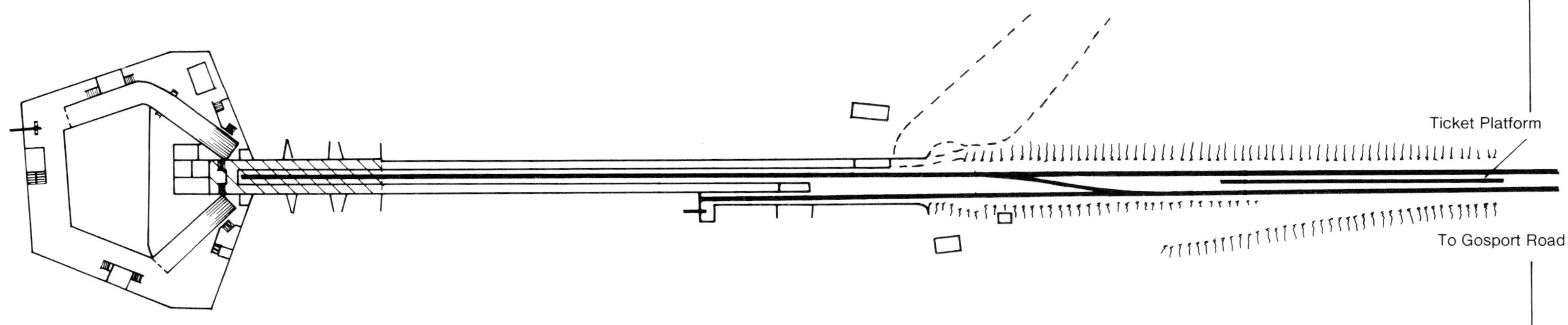

STOKES BAY SIGNALLING DIAGRAM

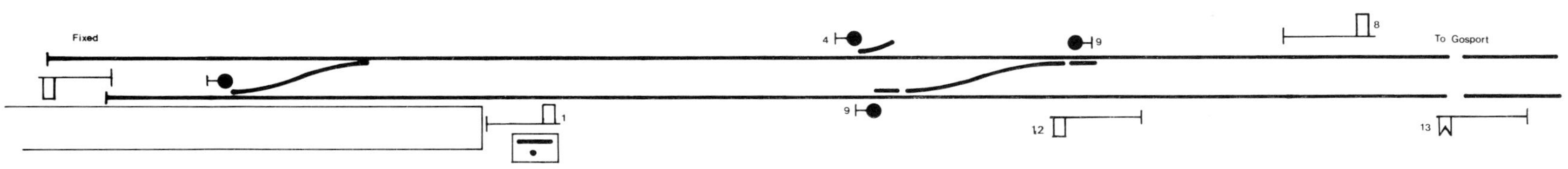

the railway Docks and Marine Department for £14. Fortunately there is no record of either vessel having to be used in an emergency.

Despite terminating but a few short feet from the sea there are no records of serious incidents, although potentially the most dangerous was on 4th February 1913 when labourer F. Jones placed a container of tar on the stove in the porters room which then boiled over. Unfortunately the result was considerable damage to the interior of the room as well as to '....nineteen panes of glass', the services of Jones were dispensed with.

The Ferry Service

From the opening of the Stokes Bay line, a ferry service had been available to passengers venturing to the Isle of Wight and whilst not strictly part of the station's history, it is worth recounting some of the details. Certain alterations were for example made to the mooring positions on the pier to assist with tying up during times of inclement weather, the biggest problems being encountered in a south-easterly gale.

As an example there were five steamer services daily in either direction in the 1880s with three on Sundays. In each case the times of train and ferry were designed to coincide where possible. Cost of the crossing was 1/- first class or 9d second and indeed the residents of nearby Alverstoke often preferred to shop at Ryde rather than Southsea. The crossing time occupied approximately 15 minutes for the 2½ mile passage. Two vessels recorded as being used on the service were the paddle steamers, *Chancellor* and *Victoria*.

A view from the back of the train shed showing what appears to be doubleheaded rails. Passengers had a short walk from this point to the ferry and no doubt were very grateful for the cover the roof afforded in bad weather conditions. *Sean Bolan*

Stokes Bay Pier 1896 with temporary platform (hatched) as used during reconstruction of the pier.

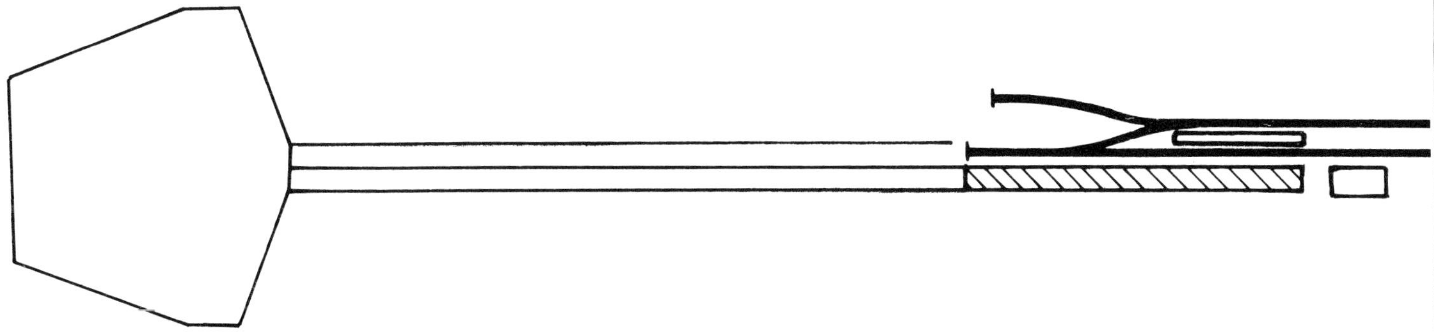

Halts on the Lee-on-the-Solent Line

A rare but poor quality view of Fort Gomer Halt c. 1910. The crossing keeper's hut is probably typical of those provided elsewhere on the branch. *Gosport Museum*

There were three halts between Fort Brockhurst and Lee, geographically from Brockhurst respectively, Privett, Browndown and Elmore. Of these the first two were original, having been opened at the same time as the Lee line in 1894. Privett Halt was itself renamed Fort Gomer Halt in October 1909 to avoid confusion with the station of the same name on the nearby Meon Valley route. Elmore Halt was a later addition and was brought into use on 11th April 1910. Being a small line, with little patronage, there was little call on the facilities the railway provided at these intermediate stopping places and, indeed, during their brief lives few changes or recordings of note have been uncovered.

The original halt facilities on the little line are thought to have been vastly different from those described above. For as previously mentioned, the Lee-on-the-Solent Company records of 1893 refer to stopping places at Pound Lane Crossing, Privett, Browndown and Elmore, the last named austensibly a passing loop. It is thought likely, however, that at the time only Elmore was actually built, having a 4 inch high platform.

Certainly by the time the line was finally opened, Elmore is not mentioned and instead Privett and Browndown are spoken of. The platforms were then of standard height. It should be stated though that the *Railway World* of July 1894 does talk of a short siding at a location called Elmore but this does not appear on maps and no other information has been found to confirm or deny the suggestion.

As was typical of minor lines of the period, a lad was employed to take charge of each halt as well as operate the adjacent level crossings. A red flag was displayed when it was required to stop a train, with the use of a red light at night. Passengers wishing to alight at either of the halts informed the guard at either Lee or Brockhurst. A small hut was provided for shelter for the lad, although no such luxury then existed for passengers!

Traffic at all the halts was small, with perhaps Browndown the busiest of them all, (often due to the use of the nearby ranges by members of the armed forces). The City of London Imperial Yeomanry were some of the most frequent visitors, but their journeys were invariably during the summer months only and therefore for the remainder of the year the inability of the railway to provide a direct link from Gosport to either Privett or Browndown led what few local residents there were to find alternative transport into the town. (The proposed curve at Fort Brockhurst would of course have allieviated this difficulty.)

The new halt at Elmore probably came about as a response to the threat of possible competition from the proposed extension of the trams to Lee. The cost of the halt is unfortunately not recorded, although details of its spartan facilities are given in the Board of Trade Inspecting Officers report. This gives its size as being 64′ x 6′ x 3′ plus ramps. Access to the platform was from a footpath. A nameboard, lamps and fencing are also mentioned, although there was at the time no shelter. The report concluded, '....sanctioned for use by rail-motor-cars on the condition that only trains of corresponding length use the platform.' The lack of a shelter was remedied soon afterwards, for in December 1910 one was provided at a cost of £32.

An old postcard view of Browndown Halt which was probably taken on a Sunday owing to the number of people present. Several interesting points are observed; the omission of 'Halt' on the station sign and the apparently concrete construction of the platform. *Lens of Sutton*

Ten years later, in February 1920, the South Western received representations from some members of the public for the provision of a shelter at Fort Gomer. Evidently it was a topic the railway had themselves been considering for at least the previous two years and therefore did not need much further persuasion. Approval was thus given for the work at a figure of £245, the cost including the provision of a booking clerk for 'special occasions'. What these may have been is not mentioned, although the Lee-on-the-Solent Company were to pay for the improvements.

Ironically, Browndown Halt was destined never to receive a shelter, instead remaining the best patronised stopping place between Brockhurst and Lee, a favourite alighting place for walkers and picnicers especially on summer Sunday afternoons.

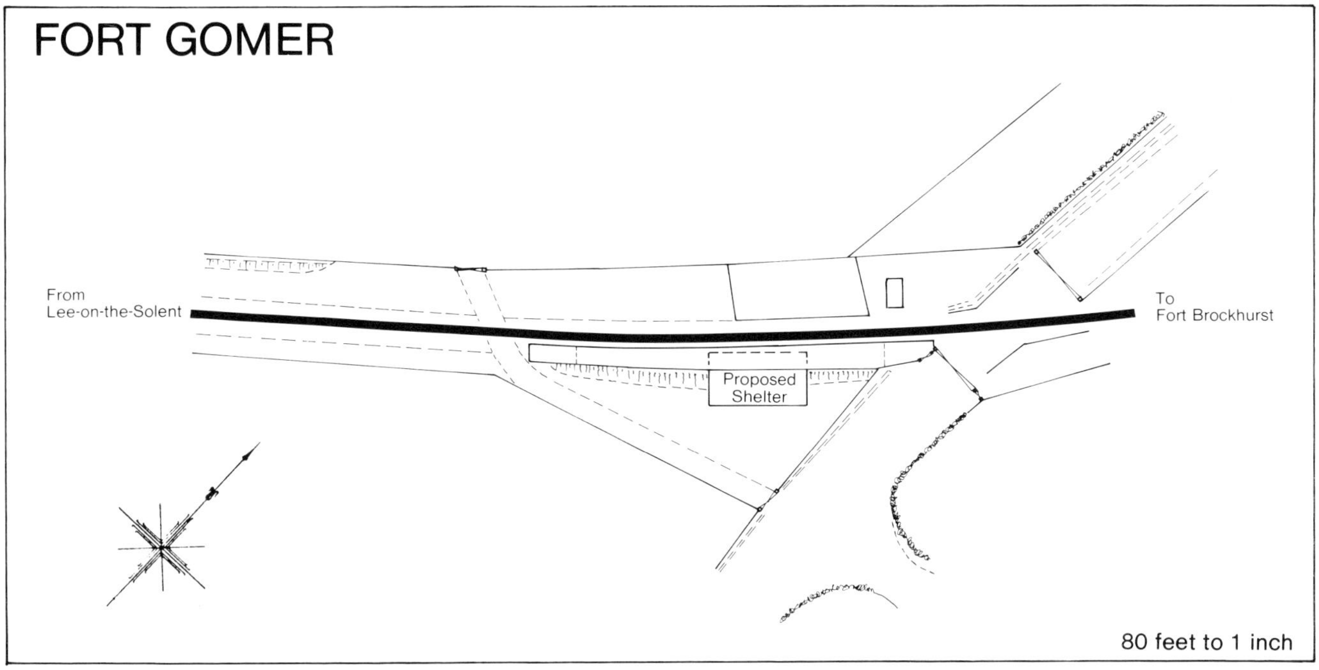

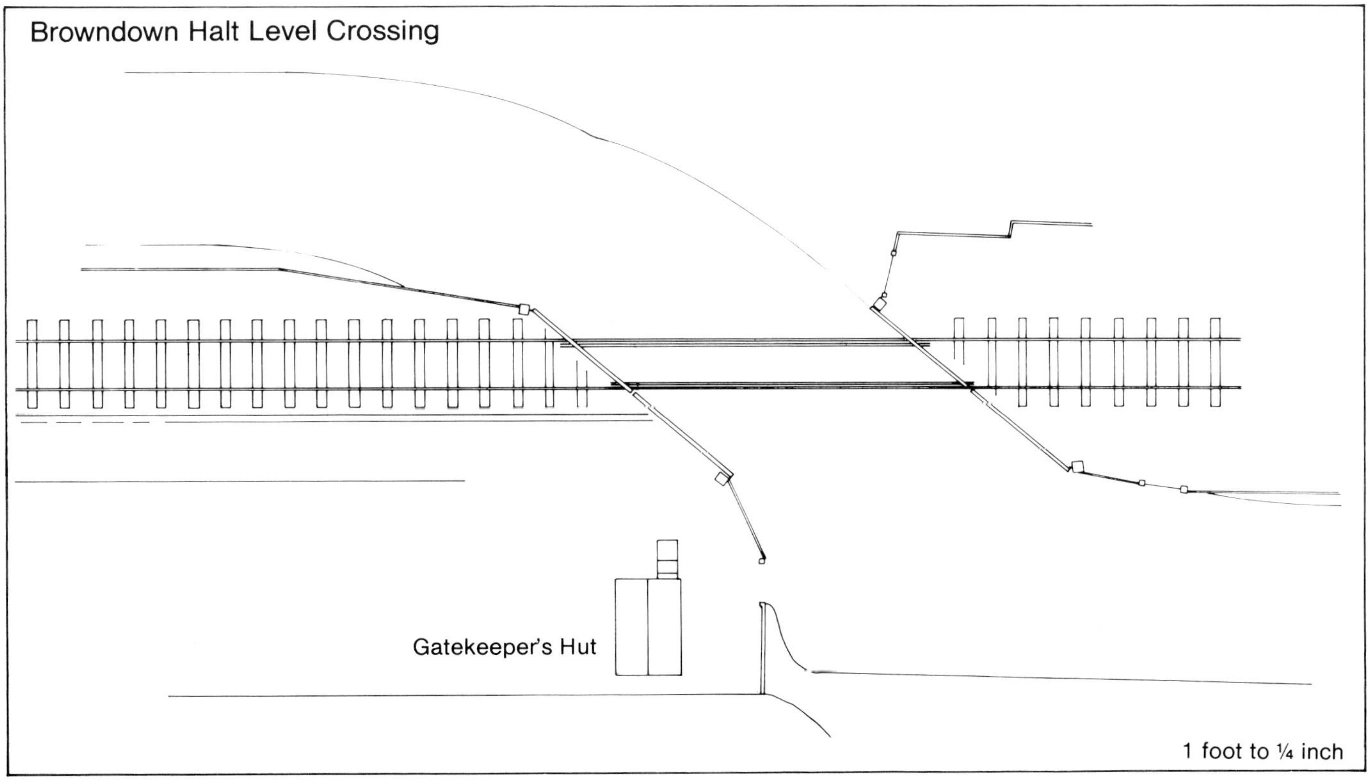

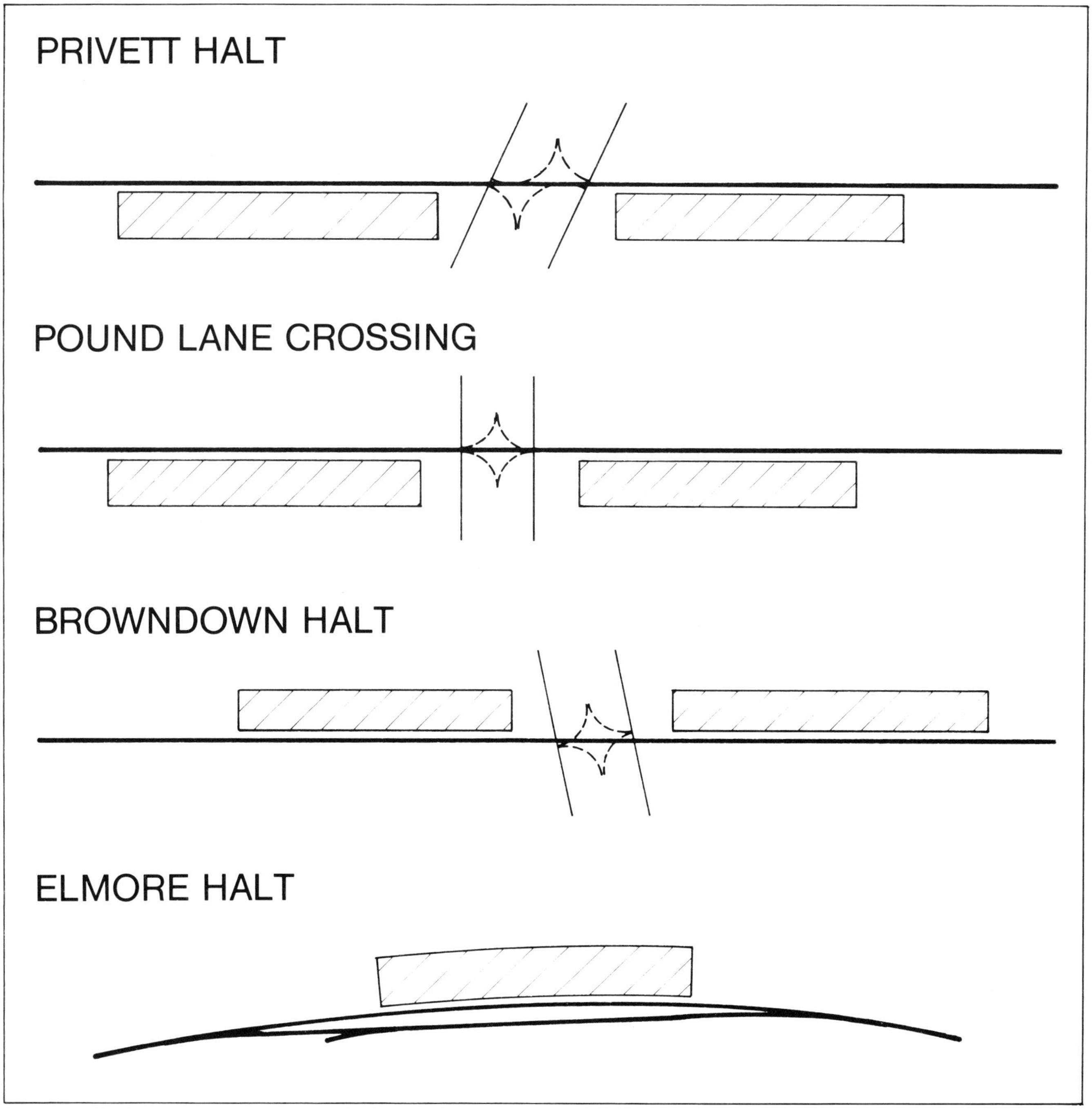

The line's path through Elmore was on a falling, curving gradient, clearly seen in this view c.1920.

Gosport Museum

Elmore Halt, probably the best patronised of all the halts on the Lee line. As can be seen, the beach was just a stones throw from the platform!

Gosport Museum

Lee-on-the-Solent

Just over three miles from its commencement the little Lee line arrived at the terminus from which its name was derived. The station was hard by the beach and approximately mid way along the 'Marine Parade' which ran the length of the shore. Adjacent was the 750 foot long pier, whilst opposite the platform and between the railway and sea stood a collection of beach huts.

In keeping with the character of the rest of the line, facilities at the station were basic. A mellow brick-built station building was at the end of the line of rails, the single platform without any form of canopy. Passenger access was from Marine Parade, by a steep slope leading down to both the beach and the railway station. To gain access to the platform it was necessary to walk through a set of double doors into a 7′ 6″ wide passage. This was paved in a diagonal pattern with 6″ square tiles. On the left were doors leading to the general waiting room and booking office, the two totally separate and both only accessible from the passageway mentioned. On the right-hand side stood some unused rooms, whilst continuing straight on, led directly into an open courtyard with further buildings on either side and the platform running away to the left.

Above: A mixed train during the last years of operation. Beach huts were a familiar feature at Lee and several were moved on to the platform itself when the line closed.
Portsmouth Evening News

Right: An early twentieth century view of Lee beach with the station terminus building on the left. Maritime admirers will no doubt notice the various clinker-built craft in the foreground. *Lens of Sutton*

LEE-ON-THE-SOLENT STATION BUILDINGS

West Elevation

Section

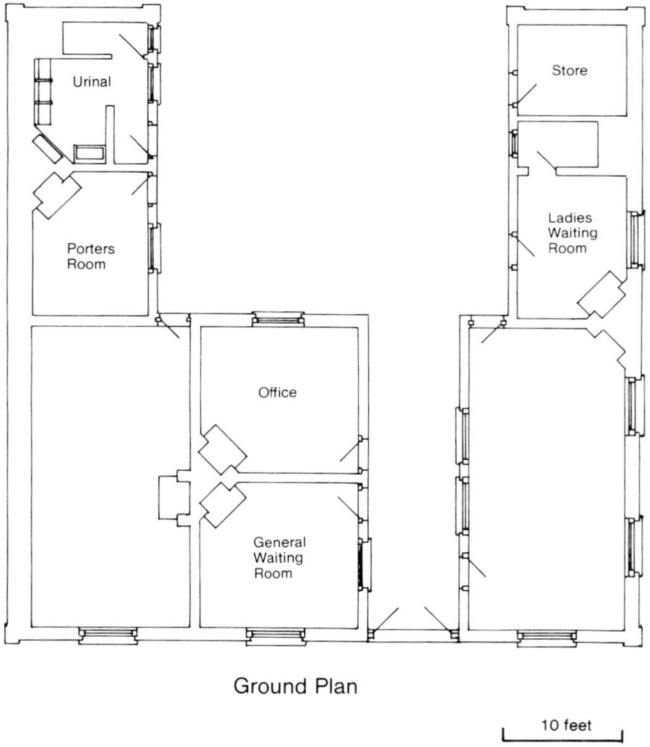

Ground Plan

Lee-on-the-Solent station building on the right-hand side of the photograph with the Pier Hotel visible in the background. *Lens of Sutton*

LEE-ON-THE-SOLENT 1910

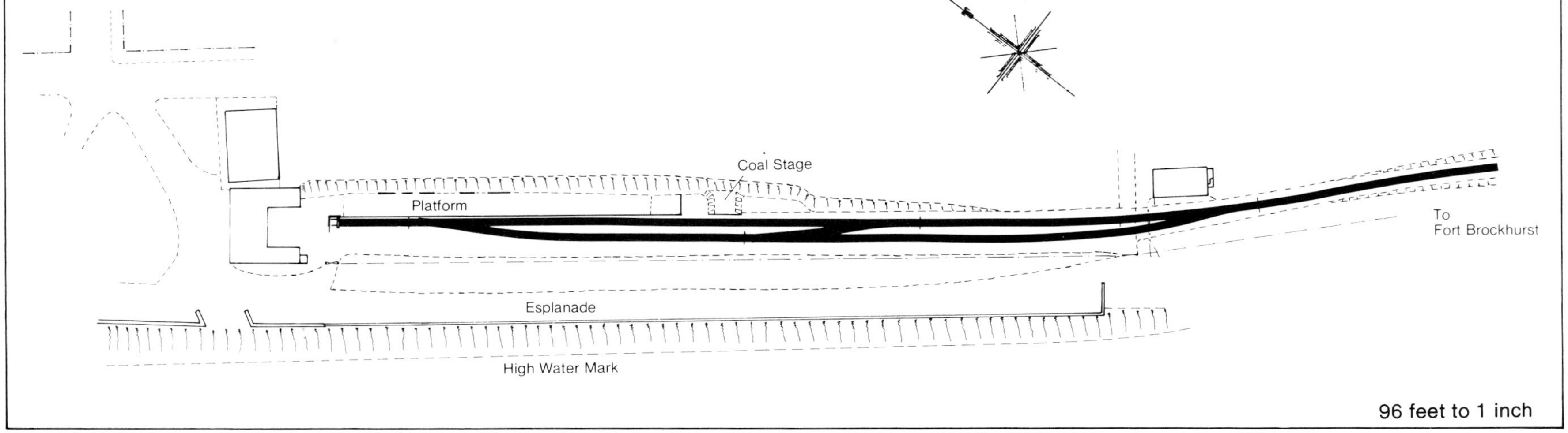

Proposed Engine Shed and Water Tank – Lee-on-the-Solent

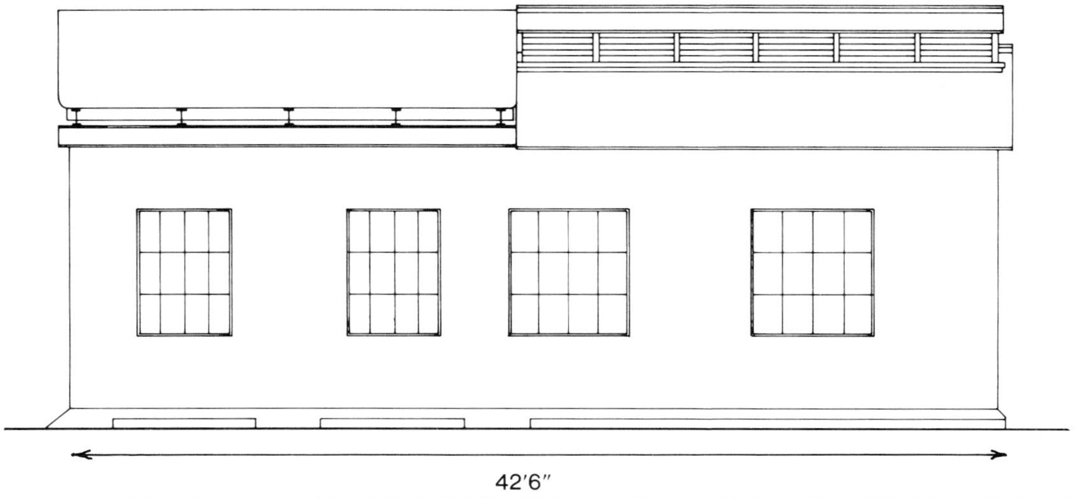

For many years LB&SCR 'Terrier' tank engines were the motive power for branch trains. Here No. B661 is at the terminus with a single coach on 14th February 1928. *H.C. Casserley*

The additional buildings mentioned provided a ladies waiting room, porters room, store and urinals, a further set of unused rooms were on the opposite side of the main building. On the platform there were also two small brick huts, one once used as an office for the stationmaster whilst the purpose of the second is uncertain, even so it is strange how these structures came to exist when there was so much unused space within the main structure.

Outside, the track layout was simple yet suitable for the traffic needs comprising a single run-round loop with crossover mid-way so that part of the loop could be used as a siding as required. There was no signalbox or fixed signals, the points on the running line being unlocked by a key on the train staff.

As built, the platform at Lee was only in the order of 90 feet long, extended either just before or shortly after opening to its later length. Despite the fact that for several years the railway operated its own services, there were only basic engine facilities at the terminus with no engine shed, although a water tank and crane supported on wooden trestles stood at the far end of the platform adjacent to which was an equally diminutive coal stage. An undated plan has, however, been found showing a proposed engine shed with replacement water tank and sited on a new stub siding at the Brockhurst end of the station. Certainly this would have made the job of the staff easier, for as one writer recounted, '....if at any time one of the staff was away a cleaner was loaned from Fratton.... the engine stood in the open and engine cleaning on the shore was no joke in the face of a south easterly gale.' The single coach, long a feature of the branch service, was stabled overnight against the buffer stop.

The directors' hopes had been that the railway would encourage development of the resort, hence the station buildings contained more space than was at first required. Consequently, from an early stage, it had been the practice to attempt to let the spare accommodation to try and gain some additional revenue, the Lee-on-the-Solent Sailing Club using one of the vacant rooms in this manner.

At the end of 1917 the Admiralty decided to establish a sea plane/searchlight depot near the station and with a build up of traffic the Sailing Club were given notice to quit. Due to an increase in parcels and other goods from the depot the extra space was then needed as storeroom accommodation. An expenditure of £26 was quoted as being necessary to render the premises suitable for their new use, all of which was against the Lee-on-the-Solent Company.

As well as the increase in parcels traffic, £1,720 was spent on some additional sidings leading into the depot. These were brought into use in 1918 and lasted just four years. No details have been found of the exact layout or the traffic handled, but the expenditure of the new work was borne by the Admiralty.

Other than these additional sidings there are few references to changes in the track layout, although also around 1918 it seems there was a general repositioning of some of the points. It would seem that changes in the traffic patterns were insufficient to warrant additional expenditure or altered facilities, although tantilisingly, as with the proposed engine shed, some undated plans exist showing two extra sidings between the platform and esplanade.

In charge of the station was a stationmaster assisted by, it is believed, two porters. The former also oversaw operations at the intermediate halts. During mid 1922 control was changed to be exercised from Brockhurst, the grade of the man in charge of Lee now being changed to that of a senior porter, although he was often still referred to as the stationmaster.

Accidents

Two accidents are recorded as occurring at the station, the first was an error which tends to show up what was a very slipshod method of working trains. On 29th September 1924 the 1.14 pm mixed train from Brockhurst arrived at Lee consisting of engine, one coach, four goods wagons and a brakevan. In between the departure of the return working at 2.05pm the wagons were gravity shunted into the siding, the engine having run-round its train previously. Unfortunately the porter in charge failed to restore the points to the normal position with the result that when the train left for Brockhurst the points were run through and damaged. Even worse, however, was that the train had left without the single line staff, the driver noticing the omission upon reaching Elmore whereupon he reversed back to Lee to be derailed at the damaged points. The driver in question was suspended whilst the porter was cautioned and later transferred to Brockhurst. Four trains were cancelled before the Eastleigh Breakdown Gang could effect rerailment.

Then on 14th February 1926 the 9.05am push-pull passenger train from Fort Brockhurst collided with the buffer stops at Lee. The leading bogie of the coach being propelled was derailed. Official records quote '....there was no complaint from either of the two passengers on board....'

Traffic and Operating

As would be expected the majority of traffic concerned passengers and of this much was squeezed into a short holiday season. Consequently summer receipts had to offset running costs for the remainder of the year.

As a matter of interest there would appear to have been a reasonable amount of freight, much of this dealt with in mixed trains and consisting of coal and building materials. With regard to the former it is likely some of the Gosport merchants would have a wagon or two of coal delivered to Lee at the time when they were undertaking deliveries in the area whilst the building materials were used in the construction of the numerous new dwellings erected at the time the line was functioning.

A party of young observers look on as the photographer captures in the background the unusual architecture of the station buildings. The 'unfinished' look about them indicates the intention to allow for expansion, which in the event did not take place. *D. Thompson*

The rival electric tramway between Gosport and Fareham seen passing Hoeford along what is now the A32 main road. *Author's collection*

Chapter Eight

Into the Twentieth Century

The new century brought with it the first winds of change to the railway. There was now the threat of real competition from the street tramways compounded with rumour that a northwards tram extension from Brockhurst to Fareham was planned.

By 1900 the horse drawn trams had been in use for some seventeen years and following some terse negotiations between the Gosport Council and the Provincial Company, agreement was reached for electrification and extension of the system. The electrified routes were to operate under the title of the 'Gosport and Fareham Tramways'. For a variety of reasons work was destined not to start until May 1905, but after this, progressed rapidly and with agreement at last reached with the War Office over certain lands in the vicinity of Bridgemary, the first tram to run the complete length of the new route between Fareham and Gosport took to the rails on 20th December 1905. The public service commenced on 24th January 1906 with cars running a 15 minute interval service between Gosport and Fareham and at 7½ minute intervals from Gosport as far as Brockhurst. A similar frequent interval service was provided on the other routes within the Gosport borough.

At Fareham the tramlines ran to within 30 feet of the railway station, Gosport station passed by the trams on their way to the tramway terminus at the Hard. This itself was close-by the main shopping centre and hence became very popular with travellers, the difficulties faced by the railway company years before, when attempting to reach the same destination, were by now overcome. The single tramway fare from Gosport to Fareham was 4d and the same as that charged on the train.

To operate the service twenty-two vehicles each with a total seating capacity for fifty-five persons were provided. Plain wooden seats were fitted, with those on the upper deck exposed to the elements. But despite such scant comfort the new service proved popular from the start and a drastic decline in passenger receipts for the railway occurred almost overnight and one which proved impossible to arrest.

At Fareham the trams provided a service within yards of the station to the town centre. Car No. 11 is seen alongside the old West Street Inn c. 1912.
Dennis Tillman collection

In connection with the electrification of another of the Gosport street tramways, that to Bury Cross, it had been necessary to extend the road bridge immediately north of Gosport Road station, this at the time near the furthermost point traversed by the trams.

With the popularity of the service proven it was understandable that the next step would be thought towards possible extension. This came in 1910 with the proposed 'Gosport and Lee-on-the-Solent Light Railway Order' and contrary to what the name suggests, was a proposed tramway extension from Bury

Cross to Lee similar to the well-known Portsmouth and Horndean Light Railway. The new line would have paralleled the existing Lee-on-the-Solent branch for part of its course and was estimated to have cost just over £9,000, although for reasons that are not completely clear, the extension was never built and instead a motor bus service began instead.

As with Lee-on-the-Solent, the South Western had been able to gain a small amount of additional revenue from Stokes Bay by leasing the pavilion on the pier to the Royal Thames Yacht Club. This, however, was only for a short period from 1913 until war brought an abrupt halt to such pastimes. Even so, plans exist of a proposed rebuilding of the pier accommodation to suit the needs of the yacht club, this in a contemporary edwardian style complete with the fineries of the period. It was not to be carried out. Other records show that for the year 1913, £78 15s 2d was received by the South Western as revenue for Tolls and Fishing Tickets from the pier.

The Alverstoke and Stokes Bay areas were also witness to developments in another form of transport in those heady days, for around 1909 the Portsmouth Aero Club had a site alongside Fort Monckton where they carried out experiments with gliders and powered flight. Additionally, there were a series of tramways, some of standard gauge, around the Haslar and what is now HMS Dolphin areas. These were mainly used for the transportation of naval goods around the named establishments and unconnected to the street tramway system. It has been suggested that there may have been a connection somewhere into the railway network to allow transfer of standard gauge wagons, this though is unconfirmed. What is certain is that there did exist an interchange between rail and tramway at or near Brockhurst (again the actual location is uncertain) and from which the occasional rail wagon of supplies was taken to various admiralty

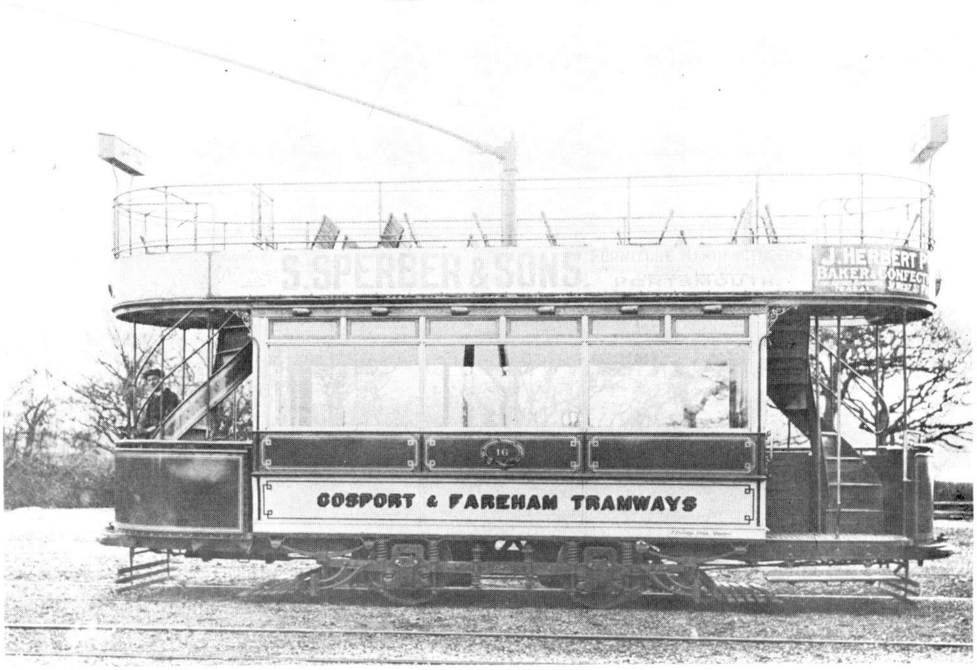

An electric tramcar from early days. No. 16 as new at the Hoeford Depot. *David Fereday Glenn collection*

Elsewhere in the Gosport area a number of street tramways connected primarily the various naval establishments. A hand propelled vehicle is seen here entering the grounds of Haslar Hospital. *Author's collection*

depots in the Hardway area. The dates between which such facilities were available are uncertain.

There were other problems too facing the railways in the years prior to World War 1, a traffic census carried out by the South Western during the winter months highlighting the lack of patronage on the Stokes Bay line. (The figures from which have unfortunately not come to light.) Consequently it became practice to suspend the ferry service from Stokes Bay during the winter months. The first time this happened was from 1st October 1902 until 30th April 1903, the period of suspension later extended to 1st June 1903. One may indeed wonder what little passenger traffic would have used the railway at this time. For the remainder of the year traffic destined to and from the Isle of Wight via Stokes Bay was brisk. One interesting working was in 1911 when a through Reading to Portsmouth working was diverted to Stokes Bay because of conjestion at the latter port.

Meanwhile, the South Western had opened their Meon Valley line between Fareham and Alton on 1st June 1903 and with it came a new route for the through trains running between Waterloo and Gosport as well as Waterloo to Stokes Bay. It was hoped this improved service would justify the expense in construction of the Meon Valley line as well as increase revenue to the Gosport area stations, but in the event this was not to be. The service provided was inconvenient and unpopular because of its slow schedules. Not surprisingly the through trains were withdrawn in 1914 and not restored. The Meon Valley service remained as a shuttle between Fareham and Alton, with some services between Gosport and Alton, to the casual observer a line laid out for speed and expansion which never came into real use.

The timetables for 1909 show no less than twenty-nine weekday movements in either direction on to the Gosport line at Fareham. A study of the workings indicate such interesting services as paths for troop trains between Fareham and Brockhurst only, a cattle train which ran on Wednesdays only and then just as required and also three Stokes Bay services which still reversed in Gosport station, these mostly with an allowance of either 5, 7 or 8 minutes to detach, run-round and then leave for Stokes Bay via Gosport Road, in itself a seemingly tight margin.

In the reverse or 'up' direction a similar number of workings are shown. One of these was a 'Special with discharged soldiers when required' leaving Brockhurst at 11.38am and due at Fareham seven minutes later. Also shown is an empty coaching stock working from Gosport to Stokes Bay arriving at 2.09pm which then formed the 2.30pm departure for Fareham. The station at Stokes Bay no doubt became a hive of activity around this time for at 2.31pm there was due a 'down' passenger working from Fareham which later returned to Gosport as empty stock. Today it seems somewhat senseless to involve empty stock working when with a slight adjustment to the timings the same set of vehicles could be used.

On Sundays in the summer of 1909 there were nine passenger workings in either direction, but with no recorded goods workings. Similarly no trains are shown on the Stokes Bay line.

The peacetime 1914 timetable shows little change, but with the addition of some local workings between Gosport and Brockhurst four times daily and twice on Sundays. Possibly a poor attempt at competition with the local tramway network.

The outbreak of war in August 1914 brought with it an immediate demand upon the railway network in the Gosport area. For with so much of the local populace dependent upon the naval and military authorities for their livelihood it is not surprising that this traffic took immediate priority over the existing services. Traffic on the main line between Fareham and Gosport and especially to and from Brockhurst, increased dramatically overnight. Clarence Yard also saw a succession of special trains, some of these with interesting wagon labels, like 'urgent, beer for troops'. Much of the yard traffic unloaded at Gosport during this period was dealt with by soldiers and mule wagons. Meanwhile, to Stokes Bay, there was an equally vast slump in traffic. This was almost wholely due to the suspension of the steamer services and which were never to reappear.

Troop trains were thus a common sight and conveyed not only men destined for the trenches but the sad remnants of those brought to be nursed back to health at the Haslar Hospital, a particularly sad time for Gosport where so many naval and military families had made their home. Consequently the families of men off to war would line the pavements and trackside as the menfolk left, the military bands striking up *'Auld Lang Syne'* as the trains departed. Equal numbers were on hand to witness the returning hospital trains, whilst at the end of the conflict there was hardly a household in the area who had not experienced the loss of a close relative.

Surprisingly the train service to Stokes Bay was provided again for the summer of 1915, but on 5th October 1915 the following minute appears in the South Western records; 'Gosport Road and Stokes Bay, falling off of passengers to and from Stokes Bay. Recommend both be closed during the period of war, the staff being released for duty elsewhere.' Closure was scheduled to take effect from 1st November 1915, but as the previous day was a Sunday the last passenger train ran on October 30th, 52 years from the time of opening.

But contrary to some reports it was not the end of rail services to the pier, for upon closure to passengers it was taken over by the Admiralty who received goods by rail, sometimes at the rate of one train daily and primarily of munitions or fuel for the bowsers established on the former promenade. The working of the line was now referred to as being a long siding shunt, unusual in as much as a double line of rails existed, the fate and consequent changes to the Stokes Bay line signalboxes are not recorded. This type of use continued at intervals right up to 1918, although by that time the frequency of trains had diminished.

Negotiations between the Admiralty and South Western over purchase of the pier began in 1919 and these indicated the railway were unwilling to reopen the line to passengers. Confirmation came in 1921/2 when completion of purchase took place for the sum of £25,000 and involving not only the pier but the railway as well to a point just south of Gosport Road station. At the time both lines of rails leading to the pier were intact. Its new use was to be as a torpedo station and degausing range, rail access stated to be no longer required.

With a return to peace in 1918 the railway network began the slow task of a reversion to normality. But of course it could never really be the same, for already the local network was contracted with the closure of the Stokes Bay line, whilst some of the railwaymen who had themselves volunteered to fight for their country were destined not to return. There was also at this time the added problem of inflation, arrears of maintenance and to an extent staff shortage, whilst again locally the tramways were a serious competitor.

The last years of the London & South Western Railway up to 31st December 1922 were a vastly different time from that of before. The new Southern Railway Company which took over from 1st January 1923, had to face up to many new challenges and conditions. One of the last acts of the old regime was in May 1922 when it was agreed to connect the stations at Brockhurst and Lee by telephone at a cost of £1 17s 6d rental per quarter together with 1½d per call made. This allowed the stationmasters post at Lee to be abolished.

'N' class 2-6-0 No. 31816 shuffles through Gosport station with a goods from Clarence Yard on 8th May 1962. *David Fereday Glenn*

Chapter Nine

Changing Fortunes

If it was possible to look forward in 1923, Gosport had but thirty more years as a railway town. In those thirty years perhaps a more drastic change was to affect the system than at any time in its past. It was now a fight for survival against seemingly increasing odds.

As recounted in a previous chapter, the Southern had also inherited the former Lee-on-the-Solent line in 1923, becoming a further worry to the accountants for at the time it could hardly be regarded as other than a loss-making concern. Perhaps surprisingly closure was not undertaken at once and instead the service continued in some respects at a better level than before. It may have been that the new owners were hopeful of an increase in fortunes, although in the event the increased train service was only to bring about an increase in difficulties for the operators.

There were two main reasons for the problems on both the main and branch lines. Firstly the local competition presented by the trams and later motor buses and secondly the inability of the railway to provide a frequent and rapid service to London and elsewhere without necessitating a change of trains. Most of these requirements were available from nearby Portsmouth. It was thus the practice for Gosport residents to undertake the short ferry ride across the water to the Harbour station in preference to the ¼ mile walk to Gosport station. The only winners in this

In the event the continuing lack of revenue forced the Southern Railway to seek economies. The decisions reached concerned not only a reduction in staff levels but also singling of the main line between Fareham and Gosport, retaining a passing loop at Brockhurst. This work was carried out in stages between late 1933 and early 1934, the revised layouts were brought into use in stages with work completed on 4th March 1934. To provide transport for some of the men involved in the track gangs, certain passenger trains made stops at the various work sites between Fareham and Gosport. At the same time as the singling of the main line, Lees Lane signalbox was reduced to the status of a ground frame.

In the meantime the track of the former Stokes Bay line had been lifted in stages so that by 1930 all that remained was part of the old 'down' line in the form of a stub extending to the boundary of Admiralty ownership just south of Gosport Road station. The former triangle though was retained for engine turning, although again only as a single line. The points at the south end were now hand-operated, Stokes Bay Junction signalbox having closed from 30th November 1924.

The 'stub' to Gosport Road was now used for stock storage. At one time there was reputedly some forty GWR 'Open-A' vehicles stabled there. The reason for this was due to the GWR Impounding Southern cattle wagons for their own use whereupon the Southern refused to return various GWR vehicles until their own stock was returned! A short while later the remains of the Stokes Bay line were cut back even further with the sale of the Gosport Road site to the local Council for £2,250. A condition of the sale was that alternative accommodation be provided for the station resident, Mr. Belbin. The way was then clear for expansion of the area, residential and commercial development over the years having now obliterated the site beyond recognition.

But it was not all contraction, for at the various annual staff dinners mention was made on one occasion, by an un-named visiting senior railway official, of possible electrification of the Meon Valley line which would in turn have been of benefit to the Gosport traffic. The same speaker also referred to the special problems confronting the Gosport railways. How much serious consideration was ever given to this proposal is un-recorded, but in the event another world war was to delay extension of the third rail network to many destinations whilst the Meon Valley and Gosport lines would never see it. At the same dinner the Mayor of Gosport, speaking as a guest, suggested a rail tunnel under the harbour to connect Gosport with Portsmouth, a valid point but one unlikely to find favour due to its high cost.

It has also been mentioned elsewhere that in 1934 the Southern had considered re-purchase and re-opening of the Stokes Bay line. In the opinion of the author this would seem most unlikely, with no evidence to confirm or disprove the suggestion it must remain mere conjecture.

But despite the lack of passenger revenue there were times when success was achieved, one example was in 1930 when following a party outing from Gosport the following letter of compliment was sent to the then stationmaster, Mr. Ford;

> 'I would like to take this opportunity of thanking you for the arrangements made for the comfort and convenience of my party. It was quite a large party, but the journey was quite a comfortable one and all spoke well of the arrangements made by you. The success of a day like this often depends on the good beginning and the good ending and the railway journey did much to ensure this.'

One may perhaps wonder as to who the party of people were and where their destination was, but on a more serious note it would seem ironic that what would probably be regarded at most stations as ordinary traffic, it warranted a special mention at Gosport. A little while later details emerge of another similar special, this time though with full details. The date was 14th June 1936 when a 3-coach train hauled by 'D1' No. 2239 took members of the St. John Ambulance Brigade from Gosport to Winchester.

Meanwhile the singling of the main line did provide an unexpected bonus in 1932, for in October of that year an agreement was signed between the Southern Railway and the Ashley Wallpaper Company for a private siding just south of Lees Lane Crossing and on the site of the former 'down' main line. This was brought into use on Tuesday 11th October.

On the left is the truncated 'down' line turned into the siding for the Ashley Wallpaper Co. The line going to the right in the distance is the northern spur of the triangle. *D. Callender*

An LB & SCR tank engine waits at Gosport on 4th August 1930.
H.F. Wheeller

Train services were in the main operated by South Western 'O2' class 0-4-4T locomotives. Several members of the class were fitted with the pulley and wire type of push-pull operation and used on the Gosport branch. From 1930 though a variation was made with the use of former LB & SCR 'D1' tank engines on push-pull trains with compressed air apparatus. Around the same time guards were removed from the Fareham – Gosport and Brockhurst – Lee services, the latter raised difficulties when it came to collecting fares from passengers joining at the intermediate halts.

But there was still some variety in the locomotives used, such as during the week commencing 25th June 1936 when 'D3' No. 2293 was reported as being in charge of certain trains. In addition the through trains to Alton remained in the hands of Adam's 'Jubilee' 0-4-2 locomotives, later replaced by 'M7' 0-4-4 tanks. Two of the six weekday services in either direction were Alton trains, a situation that had hardly changed since 1903. These were reduced still further by 1939 to one train in each direction. The goods side of operations provided a variety of motive power, primarily '700' class locomotives but also any suitable machine on a running-in turn from Eastleigh Works.

During 1932 an unusual experiment was tried on the Southern in the form of a petrol driven railcar referred to as 'The Michelin' after its French builders. Following trials in various areas it appeared on the Gosport branch where it would have seemed to be an ideal substitute for the lightly loaded trains. But what was a brave attempt at reducing running costs on minor lines was destined to failure, for whilst perhaps suitable for operating on the various light railways of France, operating restrictions in mainland Britain precluded its widespread use and after various tests, including some later on the Great Western, it quietly faded from the scene.

The passenger services themselves rarely exceeded two coaches, but well-suited to the featherweight loadings of the period, there sometimes said to be an average of only five passengers per train. The similarly patronised Lee line service started from Gosport in the morning where the coach had been stabled in the platform overnight and the locomotive in the shed. Both ran to Brockhurst where they operated the shuttle service until returning to Gosport in the evening.

Accordingly patronage was small, but there were exceptions, such as the various naval reviews and in 1929 for the Scheider Trophy Speed Trails held near Stokes Bay, Gosport station platform filled from end to end with fare paying passengers. Outgoing passenger trade too had its highlights, although apart from movement of military or naval personnel they were few and far between.

But despite the lack of patronage on the passenger side, goods traffic was heavy. The Southern recognised Gosport as primarily a goods station and accordingly the number of staff on goods work greatly outnumbered their colleagues on the passenger side. Indeed, for some years prior to 1939, freight could be said to be on the increase, various long distance through services working to Nine Elms, Feltham or Salisbury as well as local transfer trips to Fareham and Eastleigh.

Meanwhile the hopes of the Southern for the expected expansion of patronage to Lee had not materialised and instead it was recognised to be a loss maker with little likelihood of a turn round in revenue. The first stage towards closure came on 28th April 1930 when the Traffic Manager reported, '....difficulties had arisen in collecting fares of the few passengers who travel from halt to halt between the three halts on the Lee-on-the-Solent branch consequent upon the withdrawal of the guard.' It was agreed then to close all three halts as from 1st May 1930.

Unfortunately the small savings in maintenance were insufficient to forstall total closure and on 16th October 1930 the Southern Railway Traffic and Continental Committee recommended the closure of the line to passengers as from 1st January 1931. The following savings were anticipated;

Train mileage – 264 miles per week.
Immediate saving in fuel etc – £4 per week.
Saving including fuel etc when reduction of enginemen grade is effected and two junior cleaners leave service – £7 12s per week.
Ultimate fuel saving when displaced enginemen absorbed into other regular services – £11 16s weekly or £613 12s pa
Also small saving of £7 pa in maintenance of works to be dispensed with.

The final weeks of the Lee line were little different from the remainder of its life. Few passengers came to pay their last respects at a time when railway closures were regarded with quaint indifference. Accordingly all that is reported is that six men and a dog travelled on the last train, waving and singing as the service slowly wound its way from Lee to Brockhurst accompanied by the sound of exploding detonators under the wheels. Passenger traffic to Lee was now handled by bus from either Fareham or Gosport.

The goods service to Lee continued to run for a further 4½ years, although for much of this time it was on an 'as required'

Above: Posing sadly in front of their train, the staff at Lee are about to say farewell to passenger trains on the branch. Six men and a dog are recorded as being the final travellers. *Portsmouth Evening News*

Right: Four years after the withdrawal of passenger services, goods services were withdrawn. No. 2239 was in charge of the final working on 30th September 1935. *Gosport Museum*

basis, whilst diminishing in quantity all the time. The little that was carried, continued basically as before, with coal, building materials and supplies for the nearby naval air station being the main customer. Total closure came from 30th September 1935, the rails remaining in situ until around 1939 when they were lifted from Lee back as far as Gomer Halt in connection with the wartime drive for scrap metal.

The Railway At War

The anticipated declaration of war in the late summer of 1939 was recognised by the staff as likely to have an immediate effect upon the traffic in the Gosport area. These suspicions were confirmed on 31st August 1939 when with the grave decline in the international situation it was decided to mobilise plans for the evacuation of women, children and other high risk persons from Gosport, Portsmouth, Southampton and what were felt to be other likely enemy targets on the Southern system. Accordingly on 1st and 2nd September one hundred and twenty seven special trains were run from Gosport and Southampton to various 'safe' destinations in Hampshire, Dorset and Wiltshire. Many of the Gosport evacuees were destined for the stations of the Meon Valley and principally Wickham. A further eighteen trains from the three south coast towns mentioned ran on 27/28th September and again to similar destinations.

This was the period of little activity, the 'phoney war' as it was called. The situation changed from August 1940 when air raids began in what was to be in an almost regular pattern. The naval yards of Portsmouth and Gosport were understandably prime targets for the enemy.

Reputedly the first action against Gosport was in the same month, late August 1940, when bombing destroyed several buildings including the Ritz Cinema, Market House and Thorngate Hall as well as damage to an area around the Town Hall. Ironically the raids were not even targeted at Gosport but at neighbouring Portsmouth.

Then at 9.50am on 29th September 1940 a delayed action bomb was discovered ½ mile north of Brockhurst railway station and just three yards outside the railway boundary. The line was closed whilst the item was successfully dealt with.

But the actions of August and September had been but a taste of things to come and they did with avengence, the first around 8.00pm on 6th December 1940 when a high explosive bomb demolished the engine shed, blocking the running lines in the process. This destruction had in fact been foreseen by several railway staff the day before, for almost exactly 24 hours previously the same bomb had fallen into a pile of pebbles to the side of the engine shed and failed to explode. The railway staff of course notified the bomb disposal unit who attended and pronounced the weapon safe, although their mistake was in terms of damage to cost dearly. A side effect of the blast ruptured gas and water mains whilst the signalbox and a waiting train in the station were showered with pebbles. Fortunately only the signalman was reported as slightly injured. Contemporary reports are in dispute as to the state of the main line following the explosion. One report records this as only being covered in debris whilst another describes it as being blown to pieces into the nearby recreation ground. The line was cleared and services resumed in time for the 11.55am passenger departure the next day.

Later the same month the cable from a stray barrage balloon was caught in the telegraph wires just north of Brockhurst, the only casualty being the signalling equipment, pilot working was necessary until repairs could be effected. Ironically a similar situation had arisen in October 1938 in almost the same spot, when an aeroplane from the nearby naval air station brought down the wires whilst making a forced landing in a field next to the railway, again pilot working was required.

The new year, 1941, was only a few days old when on the night of 10/11th January, Gosport and Portsmouth were again victims of bombing, although on this occasion it was Portsmouth that took the brunt. But at 9.30am on the morning of the 11th, a high explosive shell fell on Gosport goods yard with another on the station approach and several incendiaries at the rear of the goods side of the station. Fortunately these latter items landed in coal wagons, although a 'stray' finished up in the stationmasters garden destroying some newly planted fruit trees! Damage to the station buildings was caused by both blast and fire with most of the windows broken and two ceilings brought down, whilst the wooden roof was damaged. In addition a wagon of cement was reported as being damaged with another of waste paper totally destroyed. In the same raid a delayed action bomb fell alongside the station approach necessitating the area being roped off and a bus service substituted until it could be rendered safe.

Two months later, on 11th March, a heavy raid was reported on both Gosport and Portsmouth. The station staff took to their shelter near the entrance to the goods yard whilst a light engine was moved to the safety of the triangle where the crew took refuge in another shelter. In the course of the action several incendiaries fell on the station, mainly on the roof, where the staff were unable to reach them. Consequently the local fire brigade were called but by the time they attended the building was blazing from end to end, low mains water pressure not helping when it came to the attempt to douse the flames. Apart from the porters room at the Fareham end of the station almost everything on the passenger side was destroyed, together with the roof. This included the stationmasters house and a wagon loaded with his effects which had been parked against the buffers. The stationmaster had been contemplating moving to the comparative safety of Wickham for some time but kept delaying a decision, it was now too late. In addition, the staff club was gutted, whilst in the booking office the furniture was charred and broken, a large table, under which had been the favourite venue for one of the clerks at times of air raids, now a smouldering piece of charcoal. Fortunately no-one had been sheltering there at the time. But in the midst of such terrifying scenes porter Gates had been able to obtain the help of the engine crew in moving some wagons from the goods side to safety, he was later commended for his actions.

With the buildings uninhabitable, provision was made for the booking and parcels departments to go to a small office on the goods platform. Temporary arrangements were later made with two small wooden huts either end of the passenger platform, one of which was taken over as an office for the stationmaster. No waiting accommodation was available for passengers, the lack of passenger traffic generally meaning there was no urgency to rectify the situation.

The same raid that had almost totally destroyed Gosport station also saw a high explosive bomb fall on the Air Ministry siding at Brockhurst derailing three wagons, although as with the raid on Gosport, without recorded injury to railway staff.

March 1941 saw three instances of explosive shells between Fareham and Brockhurst, one failing to explode, another damaging 50 feet of fencing when it went off in an adjacent field and the third, near to Wych Lane bridge, found to be an unexploded ack-ack shell considered too dangerous to attempt to defuse. Consequently it was blown up in situ. The line closed for four days whilst the resulting 40 foot crater was filled in and the necessary track repairs carried out. Understandably there were heavy delays to freight and passenger services, the passenger service eventually being suspended for one week to allow the backlog of goods to be cleared. One may wonder for a moment if thought was given to the reinstatement of the double line, certainly this would have speeded the flow on what was a very busy and at the time, strategically important route. No evidence to support this theory has, however, been found.

There was then a long gap in time before any other enemy actions, the next known damage just after midnight on 21st September 1942 when it was reported that several high explosive shells had demolished some houses near to Lees Lane Crossing, debris and rubble blocking the line. Some of the local residents took refuge in a shelter provided on the loading platform of the Ashley Private siding. The railway was reported as being blocked from 12.20am to 5.25pm that day.

This ended the recorded actions of the enemy on the Gosport line. Although it should be mentioned that at least two other undated incidents involving delayed action bombs near Brockhurst are known of, whilst in addition, damage by the enemy was caused to the former Stokes Bay Pier. Also known of is an instance of the railway being damaged near to Cambridge Road level crossing, although further details are uncertain. It is spoken of locally that at least one 'V1' may also have damaged the railway in the area.

Of the traffic dealt with during the war much was understandably military or naval in origin, as mentioned before. Large numbers of prisoners of war landed at Clarence Yard who were then marched either to a holding camp at Hardway or taken to the station for travel to Kempton Park or Moreton-in-Marsh detention camps. In the reverse direction ambulance trains were received. One of these contained a detachment of Canadian soldiers. Rumour has it that a German from a party of P.O.W.'s spat at a Canadian, whereupon a right royal battle ensued until order could be restored. There was also a regular Thursday amunition train from Gosport to Southampton, the driver of which was supposedly concerned that his train ran each week at the same time that he refused to move until he was almost unable to stand through drink. Fortunately no reference to the train being attacked is known of. Another traffic was the carriage of tanks, whilst local legend has it that Churchill, Eisenhower and Montgomery had a coach stabled in Gosport yard. But to be fair almost every station on the south coast recounts a similar story!

To handle extra traffic at Brockhurst, the remains of the Lee line at Gomer were made available to Messrs. House for unloading coal. The firm concerned shared the cost of providing the facilities required with the Southern Railway. Official records recount this expenditure as rendering the track '....suitable for use of engine and wagons.'

With peace descending upon a tired world in 1945, the railways began the slow task of catching up on the repairs the system required. These relating not only to arrears of maintenance but also to equipment damaged in the conflict. Understandably priority was given to those locations where passenger fortunes were favourable. Gosport was therefore well down the list.

Accordingly, the Southern Railway had become part of the nationalised British Railways network before authority was given to proceed with restitution. This was announced in the local press in December 1949, although the work to the station only concerning the roof. It was further said that '....eventual further rebuilding will be carried out.'

The new roof covering was erected on a steel framework partly covered with asbestos sheet. This latter feature was just over the goods side and so what few passengers remained were still totally exposed to the elements. Around the same time a small single line engine shed was erected using similar materials and on the same site as the previous structure.

The Locomotive Department

The locomotive staff at this time consisted of four crews, whilst the Brighton type of push-pull working had been replaced again and reverted to 'M7' tank operation. A locomotive of this type was stabled at Gosport for a week at a time after which it returned to Fratton for boiler wash-out. In addition to the crews there was a coalman, he worked permanent night shift during which time the locomotive was hand coaled.

For the crews themselves there were four turns of duty, the first men booking on at 5.00am ready to work the 6.10am passenger train to Fareham. These men then worked the return service to Gosport before setting off on an Alton train and changing footplates with a Guildford crew at one of the Meon Valley stations. Their replacement locomotive invariably an 'L12'. The return working was a goods, which included shunting the sidings at Funtley and Knowle before arriving back at Gosport to book off around mid-day.

The second crew out had booked on at 7.30am and they relieved some Eastleigh men on a freight which arrived at Gosport at 8.00am. The locomotive from this working was then used to shunt both the Gosport and Clarence yards, with the Eastleigh men returning 'on the cushions'. The shift finished at 4.00pm at Fareham with the Gosport driver and fireman coming home as passengers too.

Without doubt the easiest turn was that of the third set of men. The shifts worked on a rotating basis. These booked on for 11.00am and took over the locomotive from the early turn. Their work involved the passenger shuttle service from Gosport to Fareham with the last trip being the 4.30pm departure from Fareham, usually the best patronised train of the day. The tour of duty finished around 5.00pm.

A fourth set of men commenced duty at 5.00pm, their first task to shunt the yard ready for the departure of the regular 6.10pm goods. They, however, did not take this and instead worked the evening Alton passenger service as far as Droxford, changing footplates again to bring the Gosport train and locomotive back to the shed around 10.00pm. After disposing

'N' class No. 31413 provides the power this time on this service out of Clarence Yard in May 1962. *David Fereday Glenn*

of the locomotive and stock booking off time was in the order of one hour later.

The Final Years

Passenger trains still ran at basically pre-war levels, the first departure from Gosport in the morning being the workmen's train, whilst later in mid afternoon there was a Gosport to Netley working. This was designed to collect schoolchildren from Netley and return them to Fareham and Gosport.

The evening goods service along with certain of the other daytime services were worked by Eastleigh men and machines. These arrangements continued until 1951 when drastic cuts in the passenger services were made leaving only two crews at Gosport shed to work the remaining two passenger services in each direction. These left Gosport at 7.42 and 11.30am. The return workings arriving at 10.27am and 6.28pm respectively. In addition there was a 12.45pm Saturdays only service to Fareham, whilst the early morning workmen's trains in either direction were still provided but ran unadvertised. The weekday services shown in the public timetable were as through trains to and from Alton.

It was now only a matter of time before withdrawal of passenger services was announced. A taste of this had already taken place on a temporary basis in 1950 when all passenger services were taken off as a result of the then coal shortage. This time though it was to be permanent, the relevant notices being put up and passenger trains ceased to run from 8th June 1953. As this was a Monday, the last trains ran on the previous Saturday. The town of Gosport then had the distinction of being the largest town in the country without a passenger station.

With the railways in the area presenting themselves in a run-down state, there were to be occasional reminders of days past. One of these was in November 1952 when Queen Elizabeth travelled by rail to Fort Brockhurst on the occasion of a visit to the Naval Air Station at Lee. The journey started at Waterloo and consisted of 'West Country' No. 34011 at the head of a short train of Pullmans. As was normal practice, stand-by locomotives were available at Waterloo, Woking, Basingstoke, Eastleigh and Fareham, as well as two locations on the return journey, none were reported as being needed. The return working started from Fareham and ran via Havant and Guildford.

A week after closure the Coronation Naval Review was held at Spithead. No less than seventy special trains ran from Waterloo and Victoria to coastal destinations. Of these, one ran through to Gosport whilst five others are spoken of as terminating at Fareham. The question to be asked is how were these passengers transported from Fareham to either Lee or Gosport?

Fort Brockhurst's finest hour. The Queen arrives at the station on the occasion of a visit to the Naval Air Station at Lee-on-Solent.
Gosport Museum

Pictured here is one of several special trains to visit Gosport after regular passenger services had ceased. The containers mounted on trucks to the left are the railway's short-lived attempt at competition with the motor lorry. *A.E. Bennett*

But with the cessation of passenger services, goods continued to run at what can only be described as a steady rate. Some of this was the regular Bedenham siding traffic, whilst Gosport station continued to be busy. The former passenger platform was now used by the various freight services. In addition a vast council house building programme had been instigated at Bridgemary, between Fareham and Fort Brockhurst. Some of the work was carried out by foreign nationals whilst awaiting repatriation. A temporary narrow gauge railway was used to move materials around the work sites.

In addition, part of the land within the former triangle was now occupied by a firm of timber merchants. The former 'up' line on the east side of the triangle being converted to a siding for use of railway wagons loading and unloading timber. On the sides of the triangle itself were catch points protecting the main line, these the source of several derailments to locomotives whilst turning.

With passenger services no longer present and the need for such rigid controls on working relaxed, the opportunity was taken to reduce costs. The signalboxes at Gosport and Brockhurst were taken out of use with all points now converted to hand operation. This though did not prevent the occasional special passenger working, but with the various connections 'clipped' and padlocked as necessary.

But with freight a dwindling feature of the railway network it was inevitable the line would not last forever. Final closure of Gosport came in 1968 after which transfer working would only exist between Fareham and Bedenham. A situation that has lasted up to the time of writing in 1986 and appears likely to remain for some time to come.

Thus the story of Gosport's railways reaches its conclusion. The past 150 years have been witness to a slow rise in the network followed by an equivalent decline. The crumbling remains of the terminus are now in silent tribute to those years, the trains, passengers and staff, all of which have gone forever.

The posters on the gateposts were unable to attract passengers and so the station was allowed to decay, but remains a permanent reminder to those passing it that trains once were a common feature of Gosport. *National Railway Museum*

Appendix I

The Locomotives and Rolling Stock of the Lee-on-the-Solent Railway Company

As with many similar small railways elsewhere, the locomotive and rolling stock situation of the Lee-on-the-Solent line as an independent company was at times acute. The railway was run on a shoestring by the directors and accordingly with as little spent on motive power as possible. Records concerning the stock used are not always clear, and it is left to a great extent to contemporary reports to ascertain the situation.

It would seem that right from the start the railway had little intention of purchasing their own locomotives, instead relying upon the good relations established with the South Western from whom engines would be hired as required. There is though a reference in the minute books for 15th March 1894 stating that they possessed two bogie carriages, each of which was of the tramway type seating 34 passengers, 10 of which were first class. The carriages did not at first have continuous brakes, the Board of Trade requiring them to be fitted before services could commence. Both vehicles had been built by Brown Marshalls of Birmingham and similar, although larger, when compared with the 4-wheelers supplied by the same concern to the Lambourn Valley Railway. It is not known who was responsible for repairs to the vehicles, nor their subsequent fate, although it is likely that with the advent of the railmotors both were disposed of.

In addition a 4-wheeled passenger van is referred to. This is thought to have been No. 46, a 26′ fruit and brakevan built by the South Western in 1895. It is likely that it was hired to the Lee-on-the-Solent Company. Certainly other rolling stock was hired at intervals between 1894 and 1908 as required. There is no reference to any company owned goods stock.

As mentioned above, locomotives were hired to operate the service. Two well-recorded references are made to *Scott* and *Lady Portsmouth.* The first named was a 2-4-0 side tank built by the firm of George England in 1861 and in regular use on the branch until scrapped in 1909. The second engine was an 0-6-0 saddle tank built by Manning Wardell in 1862. It was first used on the branch in 1898 and is believed to have then alternated with *Scott* until December 1909.

Left: LSWR Engineers Dept. 2-4-0T *Scott.*
Gosport Museum

Below: Lee-on-the-Solent Railway composite carriage.
South Western Circle

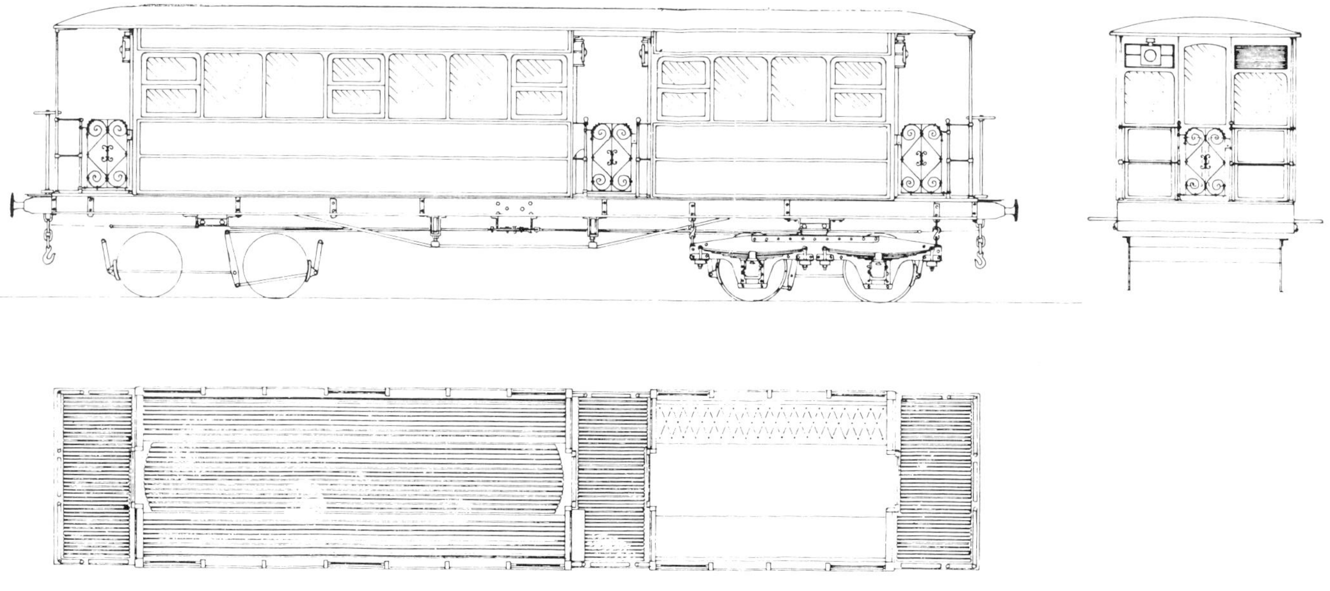

Appendix II

Locomotives of the Bedenham & Priddy's Hard Railway

Details of the contractor and locomotives used in the building of these lines are unfortunately not recorded. Similarly the subsequent changes to the layout.

No.		Type		Builder	B/No.	Date
81	Bedenhan No. 1	0-4-0F	OC	Barclay	1307	1913
82	Bedenham No. 2	0-4-0F	OC	Barclay	1389	1914
83	Bedenham No. 3	0-4-0F	OC	Barclay	1499	1916
	Bedenham No. 4	0-4-0ST	OC	Barclay	1568	1917
744		2-2-0T	OC	L.S.W.R.		1907
84	Bedenham No. 5	0-6-0F	OC	Barclay	1633	1919
88	Priddy's Hard No. 1	0-6-0F	OC	Barclay	1634	1919
85	Mining Depot No. 1	0-6-0F	OC	Barclay	1635	1919
86	Mining Depot No. 2	0-6-0F	OC	Barclay	1636	1919
87	Mining Depot No. 3	0-6-0F	OC	Barclay	1771	1922
		4wBE		G'wood	1295	1933
Yard	No. 491	4wPMR		B'ley	3028	1940
Yard	No. 205	4wDM		Hibberd	2497	1941
Yard	No. 220	0-4-0DM		Hunslet	3130	1944
Yard	No. 221	0-4-0DM		Hunslet	3131	1944
Yard	No. 215	0-4-0DM		Hunslet	3393	1946
Yard	No. 216	0-4-0DM		Hunslet	3394	1946
Yard	No. 217	0-4-0DM		Hunslet	3395	1946
Yard	No. 218	0-4-0DM		Hunslet	3396	1946
Yard	No. 219	0-4-0DM		Hunslet	3397	1946
		4wDM		Ruston	221562	1944
Yard	No. 730	0-4-0DM		Barclay	340	1940
Yard	No. 764	0-4-0DM		Ruston	319286	1953
Yard	No. 766	0-4-0DM		Ruston	414300	1957
Army	126	0-4-0DM		Barclay	361	1942
Army	127	0-4-0DM		Barclay	363	1942
Yard	No. 26653 No. 1	4wDH		Drewry	3730	1977
Yard	No. 26654 No. 2	4wDH		Drewry	3731	1977
Yard	No. 26655 No. 3	4wDH		Drewry	3732	1977
Yard	No. 26656 No. 4	4wDH		Drewry	3733	1977
		0-4-0DH		Hunslet	9045	1980
		0-4-0DH		Hunslet	9046	1980
Army	211	0-4-0DM		Vulcan	4860	1942

A Bedenham freight being collected from the exchange sidings on the Fareham-Gosport line on 31st March 1963 by 0-4-0D No. 766.

David Fereday Glenn

In addition the following eight engines are known to have worked the 2′6″ gauge lines within Priddy's Hard from 1929 until its closure in 1960.

Yard No. 320	4wBE	Greenwood	1122	1929
Yard No. 321	4wBE	Greenwood	1123	1929
Yard No. 325	4wBE	Greenwood	1163	1929
Yard No. 326	4wBE	Greenwood	1164	1929
Yard No. 327	4wBE	Greenwood	1165	1929
Yard No. 328	4wBE	Greenwood	1166	1929
	4wBE	Greenwood	1502	1937
	4wBE	Greenwood	1503	1937

There are no details available of rolling stock used solely within the depots, although it is known that standard railway owned wagons were worked within the yards by admiralty locomotives. No passenger trains are believed to have run to or from the yard.

As far as the standard gauge workings are concerned, traffic for Bedenham would be attached to 'down' Gosport freight workings at Fareham and then set off by a simple shunt move backwards into the sidings between the main line and the Brockhurst Road level crossing. The admiralty engine would then attach itself onto the rear of the train and pull it into the yard. For outward traffic the reverse procedure was adopted.

In later years a loop was installed between the main line and level crossing to assist the shunting moves.

Within the yards there were engine sheds at Bedenham and Frater with the latter used primarily for repairs. A small shed for the outstationed engine at Priddy's Hard was also provided. There was a limit on the number of wagons able to be hauled within the complex due to severe gradients.

Further details on the locomotives are to be found in *'Handbook J, Industrial Locomotives of Central Southern England'*, published by The Industrial Railway Society.

Acknowledgements

The work involved in the research and compilation of a book is a team effort from the start. Indeed it seems grossly unfair not to record the names of all those who have assisted on the front cover, for without their valued contributions none of this would have been possible. Accordingly I attempt the onerous task of compiling a list of names to whom special thanks are due. Invariably however someone will have been forgotten, my apologies if this is the case.

In alphabetical order I thank:

David Abbott – for the encouragement given, G. Allcock, Bill Bishop, Sean Bolan, Ivan Bovey, all friends at British Rail, especially Derek Clayton, Ian Coulson, John Payne, Reg Randell and Doug Stevenson, Dennis Callender, John Fairman, Walter Gilburt, David Fereday Glenn, Gosport Library, Gosport Museum, Gosport Railway Society (and especially Peter Keets), Hampshire County Record Office, Hampshire Museum, Alf Harris, John Hartfree, Chris Hawkins, H.C. Hughes – for permission to quote from his excellent paper on the Stokes Bay line, K.K. Karate, Hugh Lemon, Graham Long, Simon Mathews, Public Record Office, George Reeve, Dick Riley, friends in the Signalling Record Society, Roger Simmonds, Malcolm Snellgrove, the South Western Circle and especially Dennis Tillman, Martin Summers and especially to Peter Williamson – very grateful thanks. Also to all those photographers who have kindly allowed their work to be used.

The following books and magazines have also been consulted:

The Railway Magazine
The Southern Railway Magazine
The Railway World
The South Western Gazette
Gosport's Railway History, by G.A. Allcock.
The London and South Western Railway, by R.A. Williams.
Locomotives of the LSWR by D. Bradley.

Finally to two people without whom none of this could have been written, Roger Hardingham and Lyn Robertson – I promise I'll do some gardening next!